Artificial Intelligence

Concepts, Techniques and Applications

Yoshiaki Shirai
Computer Vision Section, Control Division
Electrotechnical Laboratory
University of Tokyo, Japan

and

Jun-ichi Tsujii
Department of Electrical Engineering
University of Kyoto, Japan

Translated by
F. R. D. Apps, Ingatestone Translations

A Wiley–Interscience Publication

JOHN WILEY & SONS
Chichester · New York · Brisbane · Toronto · Singapore

Originally published in Japanese by Iwanami Shoten, Publishers, Tokyo, 1982

Library of Congress Cataloging in Publication Data:

Shirai, Yoshiaki.
Artificial intelligence.

(Wiley series in computing)
Translation of: Jinkō chinō.
Bibliography: p. 170
Includes index.
1. Artificial intelligence. I. Tsujii, Jun'ichi.
II. Title. III. Series: Wiley series in computing.
Q335.S48613 1985 001.53'5 84–15224
ISBN 0 471 90581 X (U.S.)

British Library Cataloguing in Publication Data:

Shirai, Yoshiaki
Artificial intelligence.—(Wiley series in computing)
1. Artificial intelligence
I. Title II. Tsujii, Jun-ichi
001.53'5 Q335

ISBN 0 471 90581 X

Typeset by Activity Ltd, Salisbury, Wilts, England
Printed in Great Britain by Pitman Press Ltd., Bath, Avon

Artificial Intelligence

WILEY SERIES IN COMPUTING

Consulting Editor
Professor D. W. Barron
Computer Studies Group, Southampton University, Southampton, England

---

BEZIER · Numerical Control—Mathematics and Applications
DAVIES and BARBER · Communication Networks for Computers
BROWN · Macro Processors and Techniques for Portable Software
PAGAN · A Practical Guide to Algol 68
BIRD · Programs and Machines
OLLE · The Codasyl Approach to Data Base Management
DAVIES, BARBER, PRICE, and SOLOMONIDES · Computer Networks and their Protocols
KRONSJO · Algorithms: Their Complexity and Efficiency
RUS · Data Structures and Operating Systems
BROWN · Writing Interactive Compilers and Interpreters
HUTT · The Design of a Relational Data Base Management System
O'DONOVAN · GPSS—Simulation Made Simple
LONGBOTTOM · Computer System Reliability
AUMIAUX· The Use of Microprocessors
ATKINSON · Pascal Programming
KUPKA and WILSING · Conversational Languages
SCHMIDT · GPSS—Fortran
PUZMAN PORIZEK · Communication Control in Computer Networks
SPANIOL · Computer Arithmetic
BARRON · Pascal—The Language and its Implementation
HUNTER · The Design and Construction of Compilers
MAYOH · Problem Solving with ADA
AUMIAUX · Microprocessor Systems
CHEONG and HIRSCHHEIM · Local Area Networks
BARRON and BISHOP · Advanced Programming—A Practical Course
DAVIES and PRICE · Security for Computer Networks
SHIRAI and TSUJII · Artificial Intelligence

Contents

Preface

The object of research into artificial intelligence is to enable a computer to perform the remarkable functions that are carried out by human intelligence. Attempts to enable computers to perform symbol processing and problem-solving began to be made towards the end of the 1950s. The term *artificial intelligence* came into use with the publication of the famous paper 'Steps towards artifical intelligence' by M. Minsky of MIT in 1961. Hopes were raised in the 1960s that computers having intelligence of the level of that possessed by humans could soon be developed, by the development of systems capable of playing games such as chess, proving mathematical theorems, solving simple mathematical problems written in English, or acting as robots capable of moving building blocks. However, it was eventually realized that attaining the objective was going to be harder than anticipated because of the rapid increase in data or procedures required to achieve more than a certain level of competence. The 1970s witnessed fundamental research in all fields of artificial intelligence, such as the development of methods of searching for problem-solving, programming languages for use in artificial intelligence, adoption of the results of logic theory, techniques for interpretation of images, and methods of representing knowledge. Some practical success was achieved with symbol manipulation of mathematical equations and pattern recognition, with the result that these fields became independent of the general field of artificial intelligence. Systems which give advice using expert knowledge have also been developed in the chemical and medical fields.

In Japan, too, artificial intelligence has become an important field of information science. It is one of the fields in which the largest number of papers, such as university graduation theses, M.Sc.s, and publications of learned societies, are published. As the processing speed and memory capacity of the computer have come to approach our own, the question has come to be what the computer should be made to do and how it is to be made to do it. Research into artificial intelligence seeks methods of solving problems which do not yet have established methods for dealing with them. This is a powerful method which is useful, not only for future applications, but also for elucidating the mechanism of human intelligence.

If the field of artificial intelligence is classified by its applications, it may be subdivided into theorem-proving, games, robotics, vision, natural language processing, automatic programming, and knowledge engineering, etc. However, if the classification is made in terms of the basic methods which are used in these applications, the field may be classified into: representation of problems, searching, inference mechanisms, languages for artificial intelligence, and the representation and utilization of knowledge etc. This book is mainly concerned with fundamental concepts and various techniques in various applications of artificial intelligence, and discusses their basic principles and the methods by which they are put into practice. We hope this book will be of assistance not only in research into artificial intelligence but also in enabling application of the concepts of articial intelligence in various fields.

We are indebted to Dr. Nagao for valuable suggestions regarding the content of this book. We are also indebted to Drs. Tanaka, Yokoi, Suwa, Matsumoto, and Ikeuchi at the Electrotechnical Laboratory, Japan, for their comments on part of the draft. We should like to take this opportunity of expressing our gratitude to all concerned.

Y. SHIRAI and J. TSUJII

Chapter 1
Introduction

Progress in data processing techniques using computers has meant automation of mental work that was previously performed by human beings. In particular, once the work sequence has been established, operations such as writing up a bank ledger on withdrawal or depositing of funds, calculation of the orbit of an artificial satellite, or searching for references etc. can be performed more quickly and accurately by a computer than by people. However, the ability of computers to see an object and recognize it, or to hear, for example, the Japanese language and understand it, does not yet approach that of a human infant. The object of research into artificial intelligence is to elucidate how such mental work which does not have a determined solution sequence can be performed.

The subject matter of artificial intelligence is not therefore fixed, but changes with time. For example, about the end of the 1960s, techniques for reading handwritten letters of the alphabet or numerals were considered as belonging to the field of artificial intelligence. However, when optical character readers were developed, such techniques were no longer considered as artificial intelligence. It appears that it is the fate of artificial intelligence that, when techniques in a given field become established and put into practice, they cease to be part of artificial intelligence.

Broadly considered, artificial intelligence can be viewed from two standpoints. The first is the scientific standpoint aiming at understanding the mechanisms of human intelligence, the computer being used to provide a simulation to verify theories about intelligence. The second standpoint is the engineering one, whose object is to endow a computer with the intellectual capabilities of people. Most researchers adopt the second standpoint, aiming to make the capabilities of computers approach those of human intelligence without trying to imitate exactly the information processing steps of human beings.

However, these two approaches are closely related in that the results of scientific investigations of how people solve problems may often have considerable contributions to make towards techniques of problem-solving using computers.

The following section will introduce the various fields that form part of artificial intelligence, and the fundamental techniques that support them. The last section will then set out the scope of this book.

1.1 FIELDS OF INVESTIGATION

Research in artificial intelligence up to the present has been concerned with various problems. Effective techniques have been developed for dealing with these problems, but fresh problems have arisen. Many are common to several fields, and so have come to be regarded as the fundamentals of artificial intelligence. We shall now introduce the fields into which artificial intelligence may be sub-divided and discuss the various questions which arise in these fields, to obtain a general picture.

1.1.1 Theorem-proving

Research into methods of proving mathematical theorems using computers began in the 1950s. Success was achieved in constructing systems capable of proving theorems in elementary geometry and algebra etc. Such systems have not reached the stage of being able to prove automatically theorems which cannot be proved by human beings, but they can provide worthwhile assistance to mathematicians. Proving a theorem involves combining given axioms and rules of inference in an appropriate way to reach a conclusion. Many problems dealt with in artificial intelligence are therefore expressed in the form of proving theorems, so that solution of the problem is reduced to proving the theorem. Investigation into methods of proving theorems has had a considerable influence in other fields because it involves an investigation of the manner in which inferences are made.

1.1.2 Games

Programs capable of playing chess and draughts were developed at an early stage. Study of such games leads to the development of many techniques for searching for the best move from among the various possible moves, and these techniques have become established as methods of searching for solutions of problems. Many years of study have resulted in the development of chess-playing programs which can play the game at an advanced level. The key to success in developing a strong program is now thought to be, not so much the method of searching, but rather how knowledge about the game (openings and strategies) is adopted and used in the program.

1.1.3 Robotics

Towards the end of the 1960s attempts were made in various institutes to design 'intelligent robots' with eyes, hands and head. And, in fact, robots were produced experimentally which could recognize a simple scene and move objects about in it. The results of these studies showed that the capabilities of such robots could not be improved without further research into processing visual information (vision), control of the hands, and problem-solving etc. These various fields of research subsequently pursued separate paths, with vision becoming an independent field, and problem-solving becoming associated with theorem proving. However, the basic techniques such as problem representation and planning were developed under the stimulus of robotics. At present, robotics includes the development of sensors and controllers for controlling the position and force applied by manipulators in performing skilful operations, and the development of high-level languages suitable for describing the operating environment and giving instructions.

1.1.4 Vision

The initial objective in robot vision was the ability to recognize the shapes of simple polyhedra. However, the attempt to make a robot recognize actual building blocks showed that even this simple-sounding objective was fraught with difficulty. It is not sufficient simply to analyse variations in brightness of the input image. In fact ambiguities in the image have to be resolved utilizing the polyhedral characteristics that the object is known to have. Various techniques of image analysis were subsequently developed for recognizing people's faces, the inside of a room, outdoor scenery, aerial photographs etc. It was found that complicated scenes could be recognized if people supplied knowledge about the object to the computer beforehand in a form in which the computer could make use of it.

However, general techniques for dealing with a wide range of objects have not yet been established and methods of processing have to be devised by people in accordance with the objective.

1.1.5 Natural language processing

In order to understand the normal utterances of people, merely to construe the sentence grammatically (syntactic analysis) is not enough. Inferences concerning the meaning of the sentence, the context, and unstated facts must also be made.

If the field of discourse is restricted, information about the nouns and verbs that appear in it can be expressed in the form of a dictionary or in the form of a program. Also, the context can easily be determined. Experimental systems have been devised which are capable of understanding English or Japanese for a world which is restricted to a certain domain (building blocks, certain school textbooks, or questions and answers having a specified object in view). However, if the field of interest is widened a large number of words and contextual possibilities appear; the general knowledge and number of rules of inference necessary for language understanding increase, and it is difficult to decide which of these rules is to be applied. Problems also arise in representing such a large amount of knowledge without contradiction. Research into natural language processing has led to the development of methods of representing knowledge and high-level languages suitable for this purpose.

1.1.6 Knowledge engineering

In the diagnosis of certain types of diseases, an experienced doctor will make a diagnosis using various rules and knowledge based on experience, although the diagnostic procedure is not formalized. Studies have been aimed at the development of consultant systems, in which specialist knowledge is stored in a computer so that it can automatically give answers to questions which are not known beforehand.

A project begun in 1965 has led to the development of a system which infers the structural formula of an organic compound, given its mass analysis data. This system is called DENDRAL. It is furnished with a large number of rules for inferring the partial structure of a substance from the characteristics of the spectral data, generating possible structural formulae from the molecular formula, and predicting spectral data from the structural formula, and generates its answers by suitably combining and applying these rules. Such research has come to be known as knowledge engineering, and attempts are now being made to apply it to medicine, economics, and molecular biology, for example.

The central problems of knowledge engineering are the development of methods of representing specialist knowledge, methods of selecting and applying knowledge related to a problem, and methods of acquisition of knowledge. In particular in connection with knowledge acquisition, considerable attention has been given to developing methods whereby experts who are not very familiar with computers can enter knowledge into the computer and revise the stored knowledge when the system does not work properly. The range which can be dealt with at present is restricted; but as this range increases and the material becomes more complex, it may be expected that the central problems referred to above will become increasingly important.

1.2 SUMMARY OF THE BOOK

The previous section has explained the main fields of artificial intelligence. This book is mainly concerned with describing the basic techniques that are used in all of these fields. Specifically, Chapters 2–5 describe the representation of given problems in a strict form, methods of solving problems by combining known procedures, methods of resolving complex problems into a combination of simple problems, and methods of searching efficiently. In natural language processing and vision, it is not necessarily the case that there is a strict representation of the problem or a given set of known procedures. However, it may be expected that the techniques that are applied in these fields will include the basic techniques discussed here.

For these techniques to be used they have to be expressed as computer programs. High-level languages have been developed which are adapted for the solution of problems in artificial intelligence. Only by using these has it become possible to develop large systems. Chapter 6 describes typical computer languages for use in artificial intelligence, and explains the important mechanisms which they incorporate.

In contrast to Chapters 1–5, Chapter 7 discusses problems which can only be solved by the use of a great deal of knowledge, and explains how this knowledge is represented and used as required. The representation and use of knowledge are central topics that must be raised in artificial intelligence, though it is difficult to give an explanation simply in terms of a selection of established techniques, since this is a field in which considerable research is at present being conducted.

In Chapter 8 we mention various topics that are at present at too early a stage of development to merit a proper formal discussion, and various problems which will have to be dealt with in the future.

The object of this book is not simply to give an overview of the fields of artificial intelligence, but also to provide readers with the means to solve actual problems and to take up the challenge posed by new topics. We have tried to give a clear explanation of the essential mechanisms for solving problems, using as many examples as possible, and without burdening the text with too many strict definitions or proofs. The symbols and formulae used are in accordance with previous practice in this field. For this reason, there may be some discrepancies between chapters regarding the symbols used in corresponding formulae (capital letters and small letters may be used in different ways, for example). In particular, in applying the results of logic theory, an explanation of the exact significance of the statements and their justification has been omitted.

A list of references appears at the end of this book to facilitate further study of the various topics discussed.

Chapter 2

Representation of Problems

What type of problems constitute the subject matter of artificial intelligence? Firstly, we must exclude problems which have a well-defined method of solution. For example, finding the solution of a first-order simultaneous equation is an interesting problem in numerical computation, but does not form part of the subject matter of artificial intelligence. Here we shall deal with problems which must be solved by trial and error using a searching process.

On the other hand there are some problems which, although of great interest for artificial intelligence, will not be dealt with in this chapter. An example is problems which do not have a clear definition. Most problems of voice recognition or object recognition fall into this category. In such cases there is no objective definition of what response should be produced by what input. We shall also exclude from the scope of this chapter problems for which the knowledge or means which can be used for problem-solving are not clearly defined. Thus, problems which require a large amount of knowledge (general knowledge) or experience, such as diagnosing diseases, will also be excluded (these are dealt with in Chapters 7 and 8).

The problems to be dealt with in Chapters 2–5 have clear definitions, and the means whereby they are to be solved are also defined, though a certain degree of searching is necessary to arrive at the solutions. We shall now explain methods of representing various problems, introducing the concept of *state space*, and defining *operators* which are the means whereby a solution is obtained.

2.1 DESCRIPTION OF PROBLEMS

We shall now give examples of typical problems that we are going to deal with.

2.1.1 Reaching a goal from a starting point

First of all, we shall consider a simple maze problem. The maze is shown in Fig. 1, and the problem is to find a route from the entrance to the exit. This can be regarded as the problem of reaching the goal from the starting point through the

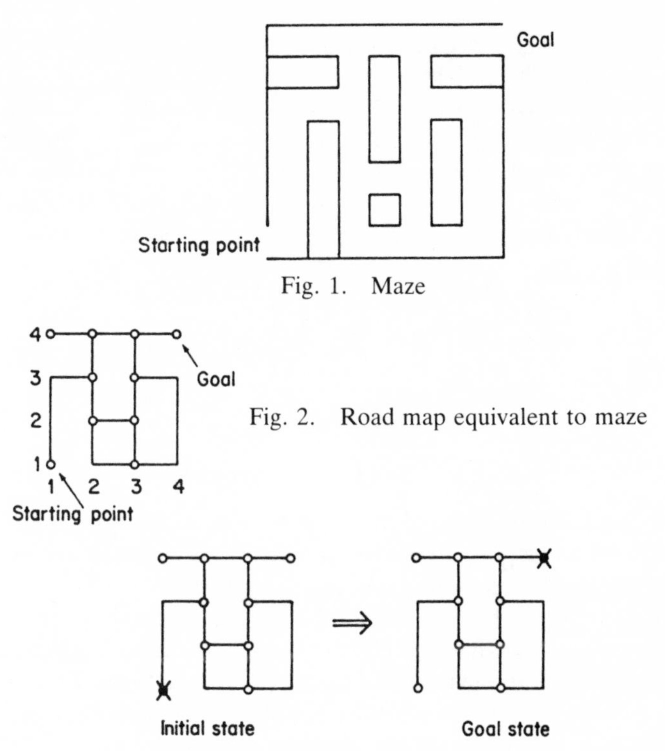

Fig. 1. Maze

Fig. 2. Road map equivalent to maze

Fig. 3. Initial state and goal state

paths shown in Fig. 2. If we represent the person travelling along these paths by X, the initial state and goal state can be defined as shown in Fig. 3. The person is free to travel along the paths and can go in any direction he chooses at intersections. There are, in all, seven intersections (mere bends in the path do not count as intersections). Positions on the paths may be expressed by coordinates as shown in Fig. 2. The first point of intersection reached as one travels away from the starting point is (2, 3). The point (1, 4) is a dead-end. The person's position changes continuously as he travels along the paths, but the only positions consideration of which is relevant to this problem are the starting point, intersections, dead-ends, and goal. These are indicated by circles in Fig. 2. The state of the problem is determined by the position of the person. The process of solving the problem involves reaching the goal state from the initial state by passing through various intermediate states.

In this problem the goal is reached by the person's movement. This movement may be defined as follows. Movement at any given position means shifting to another position that is connected to the first position by a path. Thus this movement represents a transition from one state to another. This movement is called the *operator* for solving the problem, and solving the problem may be described as finding a sequence of operators by means of which the goal can be arrived at.

We shall now consider a problem known as the '8-puzzle'. As shown in Fig. 4, this uses 8 pieces and 3×3 spaces. The object is to reach the goal state from the initial state by moving the pieces. The pieces above and below or on the left and right of an empty space can be moved into the space. The state of the problem changes every time a piece is moved. The operator for this problem is movement of a piece.

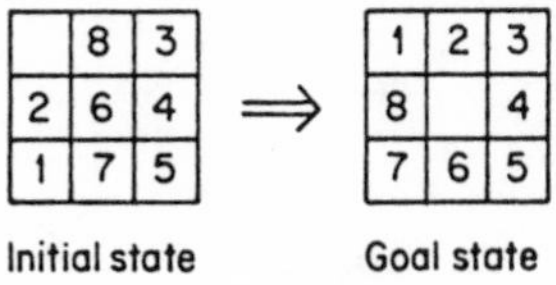

	8	3
2	6	4
1	7	5

Initial state

⇒

1	2	3
8		4
7	6	5

Goal state

Fig. 4. 8-puzzle

2.1.2 Minimum-cost problem

Consider the maze problem of Fig. 3 once more. There is not just the one route from the starting point to the goal. In fact, if we disallow routes which go over the same path twice, there are in all five possible routes. If the only requirement is to reach the goal, any of these is a correct solution.

Now suppose that moving from one position to another position is attended by a *cost* proportional to the length of the path between them. Thus the cost of any particular solution is proportional to the sum of the distance which must be travelled to reach it. The costs of the five routes can therefore be calculated. In this example, the lengths of the routes are respectively 6, 8, 10, 12. If the given problem is to find the route with minimum cost, this results in there being only one solution.

The problem of finding the solution with minimum cost when the initial state, goal state, operators and cost are defined is known as the problem of finding an *optimal solution*.

The definition of cost depends on the problem. In the case of the maze problem, the cost can be taken as the time required to reach the goal. Alternatively, if all the paths are toll roads, the problem may be defined as paying the minimum toll to reach the goal. In both cases, if a certain cost is defined for a single operator, the total cost is the sum of the costs of applying all the operators which are necessary to reach the goal. Normally, positive costs are defined for all the operators.

In the 8-puzzle, if the given problem is one of finding the least number of moves required to reach the goal, this is a problem of finding an optimal solution, with all the operators having a certain unit cost.

A well-known optimal solution problem is that of the travelling salesman. Given the road map shown in Fig. 5, the salesman starts at the starting point S and has to visit all the towns once only and then return to S. The road distances

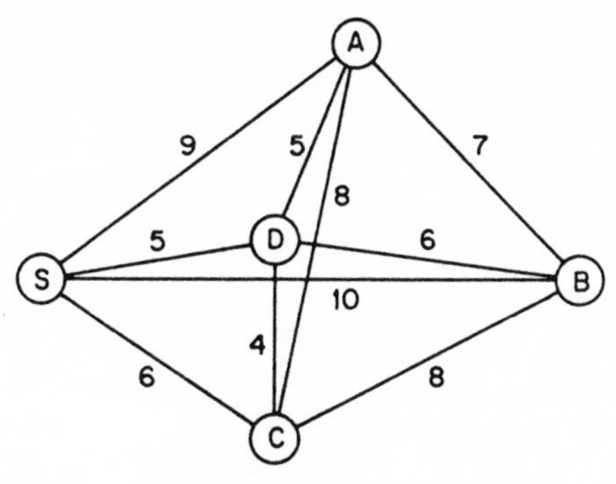

Fig. 5. Travelling salesman problem

are shown. The problem is to find the route which makes the distance travelled by the salesman a minimum. In this case, it is not suitable to use only the position of the salesman as defining the problem state. Since all the towns have to be visited once only, it is important to indicate which towns have been visited so far. This can be done by expressing the problem state in the form of the sequence of towns which have been visited by the salesman at a particular time, in the form of a list. For example, if, starting from S, the salesman has visited towns A, B, and C in that order, and is at present at town C, the problem state is (S A B C). The goal is (S X_1 X_2 X_3 X_4 S), where X_i indicates any of A, B, C, or D; and, for $i \neq j$, $X_i \neq X_j$. The number of routes that must be searched increases exponentially as the number of towns increases.

2.1.3 Games

Games can also be dealt with by these problem-solving techniques. The games which will be dealt with here are ones in which two players play alternately and no chance element is included. This includes 5-in-a-row, Go, chess, and Japanese chess (Shogi), but excludes most card games, and games using dice.

A simple example of such a game is noughts and crosses (tic-tac-toe). This is played by two players alternately making marks on a 3×3 board. If the marks made by one player are in a straight line in either the horizontal, vertical, or diagonal directions, that player has won. An example of a game is shown in Fig. 6. This figure shows the initial state of a game. The goal state cannot easily be represented by a drawing. The game of noughts and crosses is finished when one side has won or when all of the spaces are filled (this would then be a draw). If we call the players A and B, A's main goal is victory for A, and his subsidiary goal is a draw. Thus, in this example, it is more convenient to give a description of the goal state of the problem rather than to specify it concretely.

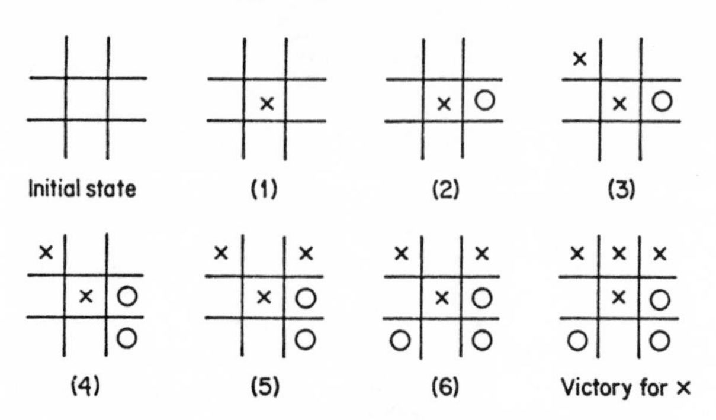

Fig. 6. Example of noughts and crosses

The state of the game may be expressed by the positions of the marks on the board and by specifying whose turn it is to play. It is easy to see from the position on the board whose turn it is to move, but this is not necessarily so in the case of games like Go or Shogi. It should also be noted that in the case of Shogi, the captured pieces, and in the case of Go, the number of captured stones, must be included when the game state is specified.

The total number of possible combinations of moves in noughts and crosses is less than 9! (The number of possible arrangements of the noughts and crosses is 9!, but, in the case of some of these, one side or the other would already have won before the board was filled). However, in most games for adults there are 10^{100} or more possible cases, so that it is impossible to examine them all. The move to be made therefore usually has to be decided by using some form of evaluation standard.

2.2 *REPRESENTATION IN STATE SPACE*

In the preceding section we have given specific examples of the terms 'state of a problem', 'operator', 'initial state', and 'goal state'. In this section we shall introduce the term *state space* to give a stricter representation of a problem. In order to formalize a problem using state space, the following must be clearly defined.

2.2.1 Description of the state

An explanation has already been given of what is meant by the state of a given problem, but it is important also to consider how such a state should be described. For example, the state of the 8-puzzle consists in the arrangement of the pieces. This can be represented in a computer in several ways. We have to choose from these the description that is most appropriate for solving the problem.

Let us suppose that the method we select for describing the state of the 8-puzzle is to specify the positions of each of the pieces. For each of the pieces numbered 1 to 8, we store the numbers of the row and column of the position of the piece. Such a description does indeed have a one-to-one correspondence with the position of the pieces, and correctly represents the state. However, it is very inconvenient for solving the problem. Firstly, it is not easy to know which pieces can be moved. Also, people find it difficult to grasp the configuration on the board.

As an alternative description let us consider a 3 × 3 matrix. The elements of the matrix will represent spaces on the board. These will be given the numerical value 0 if there is no piece in that space; or, if there is a piece in that space, will be given the number of the piece. This description is very similar to the actual state. It is also easy to see which pieces can be moved and in what direction they can be moved. Obviously this description is more suitable than the previous one.

We shall now consider the problem of the maze discussed in the previous section. In this case, the state of the problem is the position of the person. For example, when the person reaches intersection (2, 4), (2, 4) is the description of the state. Since the initially given road map does not change, it is not necessary to describe it every time the state changes.

It is not necessary that the description of the state should correspond exactly with the actual state. Let us consider a simple problem involving blocks. As shown in Fig. 7, the goal is to construct a tower on a table using blocks. In this

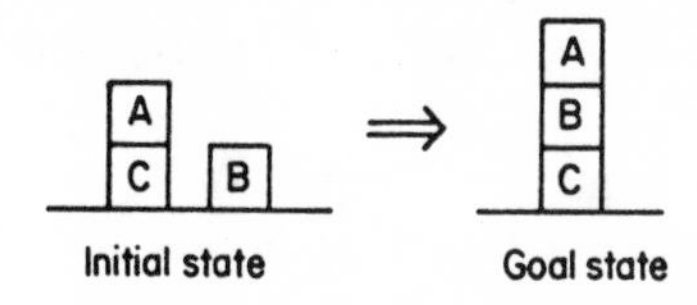

Fig. 7. Constructing a tower of building blocks

case, the position of the blocks in three-dimensional space is not very important; the problem is the relationship between the table and the other blocks. If we express the relationship that the object u is on top of the object v by ON(u, v), and represent the table by TABLE, Fig. 7 can be represented as follows:

$$\begin{array}{lcl} \text{ON(A, C)} & & \text{ON(A, B)} \\ \text{ON(B, TABLE)} & \Rightarrow & \text{ON(B, C)} \\ \text{ON(C, TABLE)} & & \text{ON(C, TABLE)} \end{array}$$

Such a *state description* is not an accurate representation of the positions of the blocks; but, if this operation were to be carried out by means of a robot equipped with an eye, it would be sufficient to specify the following steps:

1. Place A on the table.
2. Place B on the top of C.
3. Place A on top of B.

Thus a description of the state of the problem in terms of a set of relationships ON(u, v) is sufficient.

This example illustrates that the key to solving problems is to formulate a description of the states which are essential to problem-solving.

2.2.2 Operators

Operators are given directly or indirectly by the problem. An operator changes a given problem state to another problem state. An operator may therefore be regarded as a function. Its domain (input) and range (output) are the set of states. There are several ways of defining the action of an operator. One of these is to express in the form of a table the output states resulting from all possible input states. For example, an operator producing the logical sum (OR) of two logic variables x and y (which may take the values 'true' or 'false') may be defined by the outputs which it gives for the four possible inputs. However, use of such a table is not necessarily the best policy when the number of input states and output states increases. In general, an operator is defined as a process that converts a description of a given state into a description of another state.

The input state of an operator must satisfy certain conditions, called the *preconditions*. Where there are several operators, each of these will have its own preconditions.

Take the 8-puzzle as a simple example. We may define the following four operators (actually this is an 'indirect' definition of the operators):

1. Move a piece upwards.
2. Move a piece downwards.
3. Move a piece to the left.
4. Move a piece to the right.

The corresponding preconditions are as follows:

1. The space above the piece is empty.
2. The space below the piece is empty.
3. The space on the left of the piece is empty.
4. The space on the right of the piece is empty.

Since it would take too long to examine all the pieces to see whether they satisfied the preconditions, before examining the preconditions of the operators we must specify which piece is to be moved.

Accordingly, we may define the operators and preconditions as follows, if we regard movement of a piece as being equivalent to movement of the empty space:

Operators	*Preconditions*
1. Move the empty space upwards.	There is a space directly above.
2. Move the empty space downwards.	There is a space directly below.
3. Move the empty space to the left.	There is a space on the left.
4. Move the empty space to the right.	There is a space on the right.

This is a better definition because it enables the preconditions to be examined easily.

Since the operators alter the state, the state of the problem is of course determined by the way in which the operators are defined. We shall illustrate this by reference to the blocks problem referred to above. Let us suppose the robot can perform the following two types of operation:

1. Picking up x.
2. Putting x on top of y.

If these operations are chosen as problem operators, the preconditions which apply to them are as follows:

1. x is a block and there is nothing on top of it.
2. x has already been picked up, and it is true that either y is the table or that there is nothing on top of y.

To define the above, the state description used in the building block example discussed is insufficient. First of all, let us write the relationship that the robot is holding the block x as HOLD(x). Whether or not the relationship 'there is nothing on top of x' exists may be determined by checking the state descriptions to see whether there is any y which satisfies ON(y, x). However, if there is a large number of blocks, this may take some time. We therefore introduce the relationship CLEAR(x) to indicate that there is nothing on top of x. Fig. 7 may now be expressed as follows:

ON(A, C)		ON(A, B)
ON(B, TABLE)		ON(B, C)
ON(C, TABLE)	⇛	ON(C, TABLE)
CLEAR(A)		CLEAR(A)
CLEAR(B)		

The preconditions of the operators may be expressed as follows:

1. $(x \neq \text{TABLE}) \vee \text{CLEAR}(x)$
2. $\text{HOLD}(x) \vee \{y = \text{TABLE} \vee \text{CLEAR}(y)\}$

It is helpful in expressing the changes in the state produced by application of an operator to indicate the descriptions which application of the operator makes no longer true and the descriptions which are brought about by application of the operator. The former are called the *delete list* and the latter the *add list*. In this example:

1. Delete list: ON(x,$),CLEAR($x$)
   where $ indicates the object that had x on top of it before application of the operator; that is, it was an object such that ON(x,$) had been brought about.
2. Add list: HOLD(x); and, if in the delete list $\$\neq$TABLE, CLEAR($).

We omit 2. Thus, the preconditions and descriptions of the change of state produced by the operator become complicated, since it is necessary to distinguish whether the operator variable is the table or not. These can be simplified by increasing the number of operators and states.

For example, we may define the following four operators:

1. Pick up a block x from the table.
2. Pick up a block x on top of a block y.
3. Place a block x on the table top.
4. Place a block x on top of a block y.

The preconditions and effects of the operators (delete list and add list) may then be expressed in a simple way, as shown in Table 1.

TABLE 1
Description of operators

Operator number	*Precondition*	*Delete list*	*Add list*
1	CLEAR(x)	CLEAR(x),ON(x, TABLE)	HOLD(x)
2	CLEAR(x)	CLEAR(x),ON(x, y)	HOLD(x),CLEAR(y)
3	HOLD(x)	HOLD(x)	ON(x,TABLE),CLEAR(x)
4	HOLD(x) $\wedge$ CLEAR(y)	HOLD(x)	ON(x, y),CLEAR(x)

2.2.3 Constraints

Constraints are conditions that must be satisfied during the process of attaining the goal. For example, in the maze problem a condition 'the total distance travelled must not exceed a certain value' might be given. Such a condition is a constraint. Or in the problem of building a tower of blocks, there might be a constraint 'nothing must be placed on top of A'. Such a constraint need not necessarily be directly expressed.

If, in the maze of Fig. 3, there is a constraint to the effect that the total distance travelled must not be greater than L, this constraint may be included in the preconditions of the operators. The problem state is described by the person's position and the sum S of the distance he has travelled. Only a single operator is needed:

Go from (i, j) to (k, l)

If we represent the length of road between (i, j) and (k, l) by $d[(i, j),(k, l)]$, the preconditions are:

$$\begin{cases} (i, j) \text{ and } (k, l) \text{ are connected by road} \\ S + d[(i, j),(k, l)] \leqslant L \end{cases}$$

If these preconditions are used, it is possible to ensure that the distance which is moved is always less than L. Routes of length exceeding L are therefore not searched. Instead of including the distance in the constraints, an alternative would be to check, after the solution was obtained, to see whether the solution satisfied the distance constraint. This would have the advantage of saving the time needed to check the distance constraint when the operator is applied, and also that the state description could be in terms of position only. On the other hand, sometimes routes of length greater than L would be searched. Which method is best depends on the nature of the problem and the method of searching for the solution.

In the problem of building a tower using blocks, we had the precondition '$x \neq$ TABLE' for the operator 'Pick up x' which appeared first in the list of operators. We can replace this precondition by the constraint 'the table cannot be picked up'. Thus instead of checking a precondition every time the operator is applied, we check to see that HOLD(TABLE) does not appear in the states after the operator is applied.

In general constraints or preconditions are selected in such a way as to make the means of solving the problem easy to understand. In Shogi the definition of the operators is simplified if the illegal moves of doubling a pawn (nifu) or mate by dropping a pawn (uchifuzume) are treated as constraints. It is best not to be very conscious of these constraints when considering the next move, but instead to check whether the constraints are satisfied after the move has been decided.

2.2.4 Representation of state space

Solving a problem involves starting from the initial state and applying in succession operators which are applicable, while observing the constraints, to change the problem state to the goal state. During this process, the problem state passes in succession through many states. This can be regarded as movement of the problem state in state space. The problem can therefore be solved by searching state space. State space is the set of all possible states, which are related by application of the operators. In the case of a simple problem such as a maze or the 8-puzzle, the state space can be shown directly. In a complicated problem, the state space is defined indirectly by the state descriptions and the operators. That is, the state space is defined by the definitions of Sections 2.2.1–2.2.3 and the descriptions of the initial state and goal state.

The next section will be concerned with the introduction of graphs as a convenient means of representing a searching process in state space.

2.3 REPRESENTATION OF STATE SPACE BY MEANS OF A GRAPH

2.3.1 Basic concept of the graph

A graph consists of a number of nodes and edges. An edge connects two nodes. An edge may be either directed or non-directed. In the first case we have a directed graph, in the latter case a non-directed graph. A non-directed edge can be considered as an edge in both directions. If there is an edge connecting n_i and n_j, n_i and n_j are said to be adjacent, and n_i and n_j are the end-points of the edge. If there is an edge from n_i to n_j, n_i is called the start point of the edge and n_j is called the end-point of the edge. Also, n_i is termed the *parent* node and n_j the *child* node. If, in a directed sequence $(b_1, b_2, \ldots, b_m)$, for all $i < m$, the end-points of b_i coincide with the start points of b_{i+1}, such a sequence is a directed path of length m. The starting point of b_1 is the start point of the directed path, and b_m is the end-point of the directed path. If there is a series $(b_1, b_2, \ldots, b_m)$ of edges whose nodes are successively connected, in the same way as a directed path neglecting the direction of the edge, such a series is termed a *path* of length m. If there is a directional path from a node n_i to a node n_j, n_i is termed the *ancestor* node and n_j is termed the *descendant* node of n_i. If the node n is the ancestor of all the other nodes apart from n, n is termed the *root* node of the graph. If the end-nodes of an edge series constituting a path are the same, without the series containing two or more identical edges, such a path is called a *loop* (Fig. 8). A directed loop is a directional path whose starting point and end-point are the same. A graph in which there exist paths connecting all different nodes is termed a *connected graph*. A *tree* is a connected graph without loops. In a tree, there is only a single path joining any two nodes.

2.3.2 Representation by a graph

When a problem is given, the state space of the problem is defined either directly or indirectly. The state space can be represented by means of a graph. In this representation, the problem states are nodes, and two nodes are connected

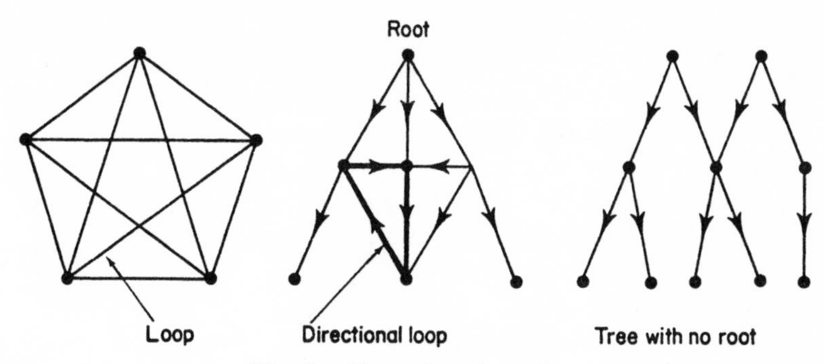

Fig. 8. Examples of graphs

by a directional edge if applying an operator will convert the state corresponding to the first node to the state corresponding to the second node. The node corresponding to the initial state is the start node, and the node corresponding to the goal state is the goal node. The procedure for obtaining the solution may be made easy to understand by giving labels to the operators corresponding to directional edges. Fig. 9 shows an example of a graphical representation of the 8-puzzle. In this graph, the label indicates where the empty space is moved to. The operator which returns to the immediately preceding state is ignored. In cases such as this in which it is always possible to return to the preceding state, the direction of the edges may be ignored.

The graph of the 8-puzzle is finite if, in constructing the graph, we do not allow further child nodes to be constructed from newly constructed nodes that represent the same state as nodes which have already been constructed. The figure shows part of a tree, but if nodes corresponding to identical nodes are represented by a single respective node, the graph would not in fact be a tree.

The state space of the maze problem of Fig. 2 may be represented by the graph shown in Fig. 10. The nodes correspond to the coordinates of each position, and the edges correspond to operators for moving between two positions. Here there is only a single operator, so the edges are not labelled. If we are finding the shortest route, and there is a constraint on the route length, it is convenient to give the edges numbers indicating the distance between the positions at their ends. Usually, where cost is part of the problem, the cost of the operator corresponding to a given edge is indicated by a number next to the edge (this is not necessary if the costs of all edges are the same). It should be noted that, since Fig. 10 contains a loop, it is not a tree.

2.4 *EXAMPLE OF REPRESENTATION OF A PROBLEM*

Up to this point we have given examples of several problems, and have demonstrated that suitable representation of a problem makes it easier to

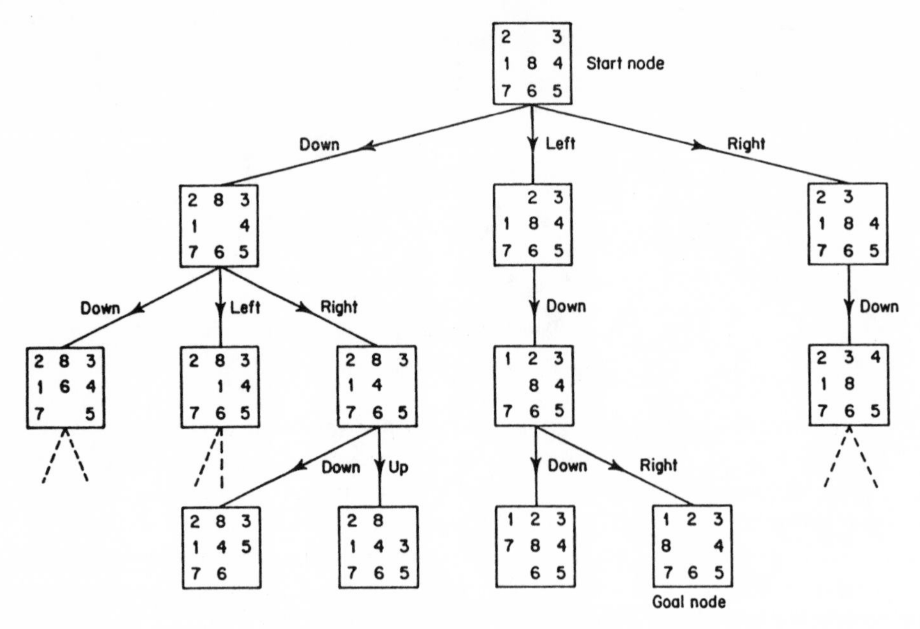

Fig. 9. Part of the 8-puzzle graph

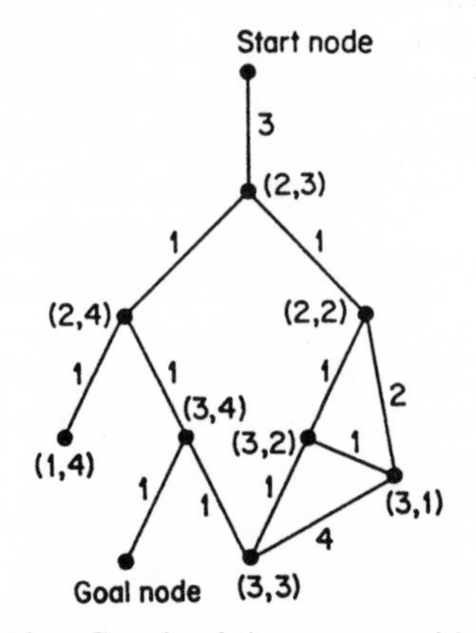

Fig. 10. Graph of the maze problems

understand and to devise methods for its solution. An outstanding example of this is the 'missionaries-and-cannibals' problem. We shall now indicate how the representation of this well-known problem may be simplified.

The problem is as follows. N missionaries and N cannibals want to cross a river by means of a boat that will hold k people. If there are ever more cannibals than missionaries on the left bank of the river, the right bank of the river or in the boat, the cannibals will eat the missionaries. Find a way of crossing the river safely.

In the case where $N = 3$ and $k = 2$, this is a very well-known problem. We shall improve the representation of the general problem in three stages.

2.4.1 Paying attention to each individual

Let us represent the missionaries by M_1, M_2, ..., N_N, the cannibals by C_1, C_2, ..., C_N, and the boat that will hold k people by b_k. Let us represent the left bank of the river by P_L, and the right bank by P_R. The initial state of the problem is as follows:

$$\begin{aligned}&AT(M_1, P_L), AT(M_2, P_L), \ldots, AT(M_N, P_L)\\&AT(C_1, P_L), AT(C_2, P_L), \ldots, AT(C_N, P_L)\\&AT(b_k, P_L)\end{aligned}$$

where $AT(x, P_L)$ means that x is at P_L. The final state is the same, with the P_L of the initial state replaced by P_R.

The operators are:

1. $x_1, x_2, \ldots, x_r$ get into the boat at $P_L (r \leqslant k)$
2. The boat goes from P_L to P_R.
3. All the crew get out of the boat at P_R.

and operators in which P_L of 1–3 above is replaced by P_R. We omit the various preconditions on the operators. $ON(x_i, b_k)$ is a state description of the state that x_i is in the boat.

Obviously, the state descriptions, operators, and constraints (concerning the number of people—omitted here) are very complicated in this representation.

2.4.2 Paying attention only to the numbers of people

In this problem the individual people are not important, so we need only consider the numbers of missionaries and cannibals. If we represent the number of missionaries and the number of cannibals on the left bank by M_L and C_L and the numbers on the right bank by M_R and C_R, and the numbers in the boat by M_b and C_b, then we may write the respective state descriptions as: $AT((M_L, C_L), P_L)$, $AT((M_R, C_R), P_R)$ and $ON((M_b, C_b), b_k)$. The operators are:

1. At P_L, M_r missionaries and C_r cannibals get into the boat.
2. The boat goes from P_L to P_R.
3. At P_R, all the crew (M_b missionaries and C_b cannibals) alight from the boat.

(Also operators in which P_L and P_R are interchanged). The preconditions and state changes of these respective operators are as follows:

1. Preconditions: AT(b_k, P_L), $M_b = 0$, $C_b = 0$, $1 \leqslant M_r + C_r \leqslant k$, $M_r \leqslant M_L$, $C_r \leqslant C_L$, $(M_r = 0) \vee (M_r \geqslant C_r)$, $(M_L - M_r = 0) \vee (M_L - M_r \geqslant C_L - C_r)$

Delete list: AT((M_L, C_L), P_L), ON((M_b, C_b), b_k)
Add list: AT(($M_L - M_r$, $C_L - C_r$), P_L), ON((M_r, C_r), b_k)

2. omitted
3. Preconditions: AT(b_k, P_R), $(M_R + M_b = 0) \vee (M_r + M_b \geqslant C_r + C_b)$
   Delete list: AT((M_R, C_R), P_R), ON((M_b, C_b), b_k)
   Add list: AT(($M_R + M_b$, $C_R + C_b$), P_R), ON((0, 0), b_k)

Obviously this has greatly simplified the state description and facilitated the definition of the operator preconditions and state change.

2.4.3 Removal of redundant representations

Let us consider the constraint on numbers of people when the boat moves from the left bank to the right bank (this is included in the preconditions in Section 2.4.2). On the left bank, the following condition is always satisfied:

$$(M_L = 0) \vee (M_L = C_L) \vee (M_L > C_L) \tag{1}$$

A similar condition is satisfied on the right bank. When the boat is at either bank, $M_b = C_b = 0$, so on the right bank:

$$(N - M_L = 0) \vee (N - M_L = N - C_L) \vee (N - M_L > N - C_L) \tag{2}$$

The following relationship may be deduced from equations (1) and (2):

$$(M_L = 0) \vee (M_L = N) \vee (M_L = C_L) \tag{3}$$

If the people are to be transported while maintaining this relationship, the changes in numbers of people shown in Fig. 11 are possible. For these, the conditions shown in the figure on the respective numbers of people M_r and C_r are necessary. Clearly the constraint on the number of people in the boat is satisfied in each case. A similar situation obtains in the case of the right bank. Obviously then, if the constraints on both banks are satisfied, the boat constraint is not necessary. Furthermore, it can be seen that if equation (3) holds on the left bank, a similar condition must hold on the right bank. Taking all these things into account, Fig. 11 is obtained.

If we take the state as being represented only by the number of people on the left bank and the position of the boat, we may write the state description as (M_L, C_L, p). In this description, if the boat is on the left bank, $p = 1$, and if it is on the right bank, $p = 0$. The problem is now defined as:

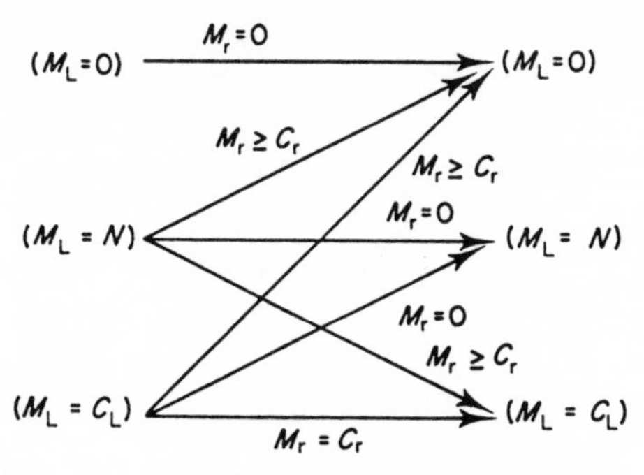

Fig. 11. Possible changes in number of people on left bank

$$(N, N, 1) \Rightarrow (0, 0, 0)$$

The operators are movement of the boat from the left bank to the right bank, and the reverse. These can be represented in a very simple way:

1. Preconditions: $p = 1, 1 \leqslant M_r + C_r \leqslant k$,
   $(M_L - M_r = 0) \vee (M_L - M_r = N) \vee (M_L - M_r = C_L - C_r)$
   Delete list: $(M_L, C_L, 1)$
   Add list: $(M_L - M_r, C_L - C_r, 0)$
2. Preconditions: $p = 0, 1 \leqslant M_r + C_r \leqslant k$,
   $(M_L + M_r = 0) \vee (M_L + M_r = N) \vee (M_L + M_r = C_L + C_r)$
   Delete list: $(M_L, C_L, 0)$
   Add list: $(M_L + M_r, C_L + C_r, 1)$

Since the number of possible states has been cut, the state space is made smaller, and so less searching is required. This shows how solution of a problem is made easier by choosing a representation for the problem that makes effective use of the constraints.

Chapter 3

Searching Techniques

In the preceding chapter we showed how, if a problem could be represented in state space, solving the problem could be reduced to a searching process performed in state space. The searching technique must be suitably chosen for the problem. We shall now describe basic searching techniques. The simplest technique is to select operators by a trial-and-error process; but if finding a solution must be guaranteed, there is no alternative but to undertake a systematic search of the state space. If the minimum cost solution is to be found, it is more efficient to carry out a search that is specifically adapted to this end. If the cost of reaching the goal can be predicted beforehand, the searching time can be cut by effective use of this information. We shall now explain suitable algorithms for performing such searches.

3.1 TRIAL-AND-ERROR SEARCHING

The simplest method of searching is to keep on applying arbitrarily selected applicable operators until the goal state is reached. This method is represented by the procedure given below. (In the procedures given in this book, the remarks after the semicolons are comments on the various steps.)

Procedure *search*

1. Put *state* in initial state.
2. **While** *state* $\neq$ goal state **do**.
3. **Begin**.
4. Select operator applicable to *state* and make this *operator*.
5. *State* := *operator* (*state*); this applies the *operator* to *state* to produce a new state, which then becomes *state*.
6. **End**.

In step 4, an operator is randomly selected from the several applicable operators. This makes the procedure stochastic, with no guarantee that the goal will necessarily be reached. However, this trial-and-error search procedure would be capable of solving a simple problem such as that of Fig. 3. (This procedure is simply a procedure for reaching the goal state. If the path is to be found, the successive changes of state have to be recorded in the memory.)

The following different methods of searching are obtained by using various methods for selecting the operator in step 4.

3.2 SEARCHING BY SYSTEMATIC NODE EXAMINATION

In this method the state space of the problem is examined in some predetermined sequence until the goal state is found. If the state space is finite and a solution exists, this method will always find it. The method may be subdivided into *vertical* and *horizontal* search methods, depending on the order in which the search is carried out when the state space is represented graphically.

3.2.1 Vertical search

First, consider searching when the state space of the problem is represented by a tree. In the vertical search method, the nodes which are deepest in the tree are searched first. It is therefore also known as a *depth-first* search. For instance, in Fig. 12 the order of the search starting from the starting node S until the goal node G is found is A, C, D, B, E, H, G.

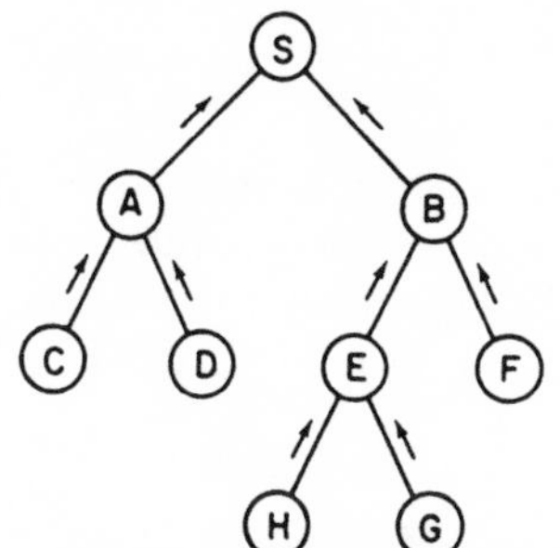

Fig. 12. Example of a tree, with arrows representing pointers

If the node n that is being checked is not G, if n has a child node, the child node is examined next. If n has no child node, the program returns to the parent node m of n, and examines the next child node after n below m. The path to the goal node can be obtained by arranging that the parent node of each node is stored. This may be done by giving each node a pointer to its parent node. In Fig. 12 the pointers are indicated by arrows. When the goal node is found, the path can be obtained by following the pointers back to the starting node. The solution is found by following this sequence of pointers backwards.

We shall now give an algorithm for the depth-first search. We shall call the process of finding the child nodes of a node n by applying to it all the applicable operators *expanding* n. Normally, in solving a problem, instead of generating the tree beforehand and then searching it, the tree is generated by expanding the nodes as the search proceeds. The tree is defined indirectly by the state space representation. Nodes which have already been generated but not yet expanded are entered in a list called the *open* list.

Procedure *tree search*

1. Put the start node in *open*.
2. *LOOP*: **if** *open* = empty **then** *exit* (*fail*); if all the nodes have been examined, the search terminates, the result of the search being failure.
3. *n* := *first* (*open*); fetch the initial element from *open*.
4. **If** *goal* (*n*) **then** *exit* (*success*); if *n* is the goal node, the search has succeeded, and is terminated.
5. *Remove* (*n, open*); *n* is removed from *open*.
6. Expand *n*, put all the child nodes at the head of *open*, and attach pointers from the child nodes to *n*.
7. **Goto** *LOOP*.

Goal (*n*) in step 4 is a logic function which is true if *n* is the goal node, but is false otherwise. In step 6, if *n* has no child nodes, the computer does nothing, but simply goes to the next step. If it has several child nodes, these may be entered at the head of *open* in any order; but if one of these is the goal it is more efficient to put this at the head.

When this procedure is applied to the tree of Fig. 12 the open list changes as follows:

$$(S) \rightarrow (A\ B) \rightarrow (C\ D\ B) \rightarrow (D\ B) \rightarrow (B) \rightarrow (E\ F) \rightarrow (H\ G\ F) \rightarrow (G\ F)$$

If a graph is being searched, any child nodes that are generated in step 6 which have already been examined do not need to be entered in *open*. We therefore put nodes which have already been expanded in a list called the *closed* list (*closed* is initially empty). For a depth-first search, steps 5 and 6 may be altered as follows:

5. *Remove* (*n, open*)
   Add (*n, closed*); *n* is added to *closed*.
6. Expand *n* to generate all the child nodes. Any child nodes which do not appear in *open* or *closed* are put at the head of *open*, and given pointers to *n*.

In the procedure described above, for simplicity, all the nodes are expanded at once. However, it is usually more efficient to generate the child nodes one at a time.

3.2.2 Horizontal search

In a horizontal (breadth-first) search, the shallowest nodes of the tree are searched first. In the example of Fig. 12 the order of searching would be A, B, C,

D, E, F, H, G. In the case of a graph, the nodes are examined in order starting from those nearest to the start node. An algorithm for a breadth-first search of a graph may be written as follows:

Procedure breadth-first search
1. Put the start node in *open*.
2. *LOOP*: **if** *open* = *empty* **then** *exit* (*fail*).
3. $n := first\ (open)$.
4. **If** *goal* (n) **then** *exit* (*success*).
5. *Remove* (*n, open*).
6. *Add* (*n, closed*).
7. Expand n and generate all the child nodes. Any of these child nodes that are not included in *open* or *closed* are to be put at the bottom of *open*, and given pointers to n.
8. **Goto** *LOOP*.

This procedure is exactly the same as the depth-first search except for step 6. Whether the search is a depth-first search or a breadth-first search is determined by whether newly generated nodes are placed at the head or at the bottom of *open*.

In general, if the goal node is at a deep location in the tree, a depth-first search is most suitable, while if it is at a shallow location, a breadth-first search is more suitable. However, if the depth of the tree and the average number of branches from a single node (effective branching factor) are beyond a certain level, whichever method is adopted, the number of nodes of the tree which must be searched becomes very large indeed.

3.3 SEARCHING FOR THE OPTIMUM SOLUTION

Consider the problem of finding the minimum cost path in a case in which the state space is represented by a graph and the cost of the operators corresponding to each edge is given. All the costs are positive.

First, for the sake of simplicity, let us deal with searching a tree. Fig. 13 shows an example of a tree to be searched (the goal nodes are G1 and G2). The policy of the search is to examine first those nodes which have minimum cost from the start node. In the example shown in the figure, nodes A and B are generated by expanding the start node S. Since the B is the node of least cost from S, we examine B. Since B is not the goal node, B is expanded, generating E and F. Now, of A, E, and F, the cost to E is the least; so we examine E and expand it. This results in nodes H and G1 being obtained. The solution is obtained by repeating this procedure.

An algorithm is given below. We represent the cost of the optimum path from the start node to the node n by g(n), and the cost of the best path found so far during the search as $\hat{g}(n)$; that is, $\hat{g}(n)$ is the inferred value of $g(n)$. If a tree

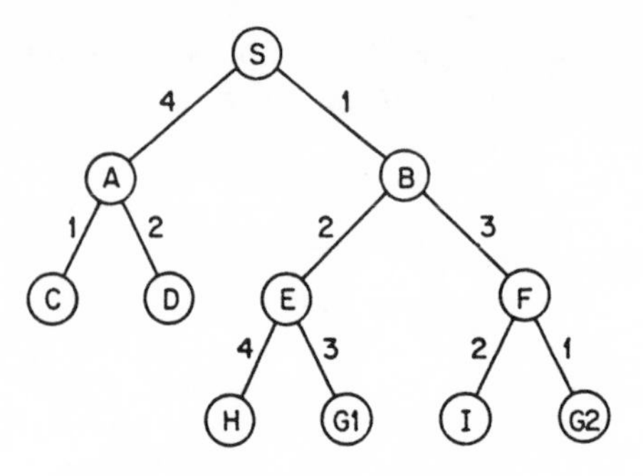

Fig. 13. Tree with defined costs

is being searched, $g(n) = \hat{g}(n)$.

Procedure *tree-search-1*

1. Put the starting point S into *open*. $\hat{g}(\mathrm{S}) := 0$.
2. *LOOP*: **if** *open* = empty **then** *exit* (*fail*).
3. $n :=$ *first* (*open*).
4. **If** *goal* (n) **then** *exit* (*success*).
5. *Remove* (n, *open*).
6. Expand n, calculate $\hat{g}(n_i)$ for all the child nodes n_i, and attach pointers from n_i to n. Put all the child nodes in *open*, and arrange all the nodes in *open* in order of least cost.
7. **Goto** *LOOP*.

In step 6, if we take the cost of the edge connecting nodes n and n_i as $c(n, n_i)$, the cost of the child node n_i may be calculated by $\hat{g}(n_i) = \hat{g}(n) + c(n, n_i)$. When the tree of Fig. 13 is searched by this procedure, the *open* list changes as follows (the cost from S to each node is given in brackets):

(S(0)) → (B(1) A(4)) → (E(3) A(4) F(4)) → (A(4) F(4) G1(6) H(7)) → (F(4) C(5) G1(6) D(6) H(7)) → (G2(5) C(5) G1(6) D(6) I(6) H(7))

In step 6, nodes whose cost is the same may be listed in any order, but, in the above example, the goal node is listed first, followed by the shallowest nodes.

It should be noted that, even if the goal node is generated in step 6, it does not follow that the optimum solution has been obtained. If we examine the changes in the *open* list in this example, we can see that there is an F in the penultimate list which has a cost less than G1; and the cost of G3, which can be reached from F, is less than the cost of G1. We can be sure that, in the open list, the cost of a goal node other than G2 must be equal to or greater than the cost of G2. We can see that this procedure enables the number of nodes that must be examined to be cut. For example, in the tree of Fig. 13, even if the nodes C, D, H, and I have many child nodes, the procedure terminates before these are examined. Often there is a great saving in time compared with a method of searching in which the optimum solution is found by examining all the nodes and then finding the complete cost to the goal node. If the costs of the edges of the graph are constant, tree-search-1 corresponds to a breadth-first search.

We shall now generalize the problem to that of finding the optimum solution in a graph. This means that the child nodes that are expanded in step 6 of tree-search-1 may include nodes that were already generated when other nodes were expanded. In the previous section, the same node would not be expanded twice, so nodes that were generated for the second or subsequent times were not put into *open*. However, if the cost of these nodes that were generated twice or more is small enough, their path may be a part of the solution. Step 5 of tree-search-1 must therefore be altered as follows:

5. *Remove* (*n*, *open*).
   Add (*n*, *closed*).
6. Expand n and calculate the cost $\hat{g}(n, n_i)$ to n_i through n from S for all child nodes n_i. Put nodes that are not found in *open* or *closed* in *open* and make $\hat{g}(n_i) = \hat{g}(\mathrm{n}, n_i)$, setting pointers to n from n_i. For nodes that are already contained in *open*, compare $\hat{g}(n_i)$ with $\hat{g}(n, n_i)$ before expanding n, and if $\hat{g}(n, n_i)$ is smaller make $\hat{g}(n_i) = \hat{g}(n, n_i)$ and set a pointer from n_i to n. Do nothing for the other child nodes. Lastly, list the nodes in *open* in order of lowest cost.

This procedure is termed *optimal-search*. If the cost of the edges of the graph is constant, optimal-search corresponds to a breadth-first search. The successive changes in the *open* list when this procedure is applied to the graph of Fig. 14 are as follows:

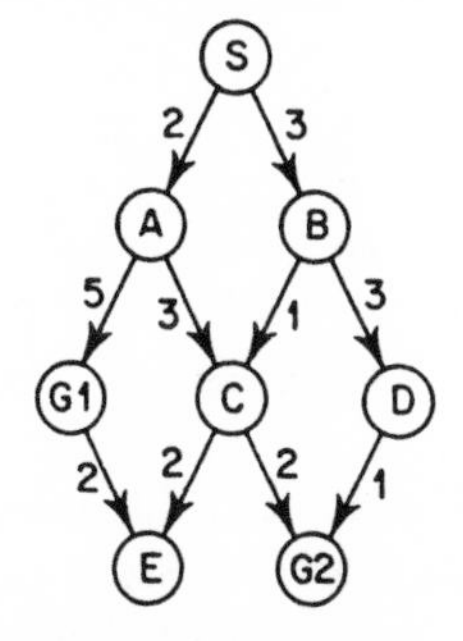

Fig. 14. Graph in which cost is defined

(S(0)) → (A(2) B(3)) → (B(3) C(5) G1(7)) → (C(4) D(6) G1(7))
→ (G2(6) (D(6) E(6) G1(7))

It can be seen that the inferred value of the cost of node C alters while the procedure is being carried out.

It is easy to show that if the graph is finite and there is a path from the start node to the goal node, the solution will always be obtained by an optimal-search procedure.

Theorem 1

If a graph is finite, and there exists a path from the start node to the goal node, the optimum-search procedure will always be successful.

Proof. Assume that the optimal-search procedure fails. In such a case, the *open* list would be empty, while the *closed* list would contain all the nodes searched up to that point. Since the graph has a solution, take any solution and express the series of nodes along the path of this solution by $(n_0, n_1, \ldots, n_m)$ (n_0 is S, n_m is the goal node). Tracing this series backwards, we find a node n_i that was entered in *closed*. That is, $n_i \in$ *closed*, and $n_i + 1 \notin$ *closed*. There must be such a node n_i (the reason being, $n_0 \in$ *closed*, $n_m \notin$ *closed*). Since n_i is included in *closed*, it will have been expanded at step 6. If this happens, $n_i + 1$ must have been put into *open* once. So n_{i+1} must have been examined before *open* became empty. If n_{i+1} is the goal node, the search succeeded. If it is not the goal node, n_{i+1} is put into *closed*. In either case, the assumption that the search failed is contradicted. Since the graph is finite, if the search does not fail it must succeed. (Theorem 5 shows that the search will succeed if there is a finite solution path, even if the graph is not finite.) The reason why a child node n_i that is already in *closed* is not again put into *open* is that $\hat{g}(n, n_i)$ cannot be less than $\hat{g}(n_i)$, so a path passing through n need not be considered as a candidate for the solution. This is shown by the following theorem.

Theorem 2

At the time-point when the node n is expanded, optimal-search has already found the optimum path up to n ($\hat{g}(n) = g(n)$).

Proof. We can say that the inferred value of the cost of the node at the head of the *open* list decreases as the search proceeds. In optimal-search, every time the computer goes through the loop, the node n at the head of *open* is removed; and if there are child nodes of n, a partial set of the set of these child nodes is added, and, if required, the cost for the partial set of the nodes in *open* is altered. For the child node n_i of n, we have $\hat{g}(n, n_i) = \hat{g}(n) + c(n, n_i)$. Since $c(n, n_i) > 0, \hat{g}(n, n_i) > \hat{g}(n)$.

Consequently, both the cost of the nodes which have been added to *open* and of nodes whose costs have been altered are greater than $\hat{g}(n)$. That is, the cost of the node at the head of *open* does not decrease. Since costs will not be found in the future that are smaller than $\hat{g}(n)$ of the node n at the head of *open*, $\hat{g}(n)$ is the cost of the optimum path.

3.4 SEARCHING WHEN COST UP TO GOAL CAN BE PREDICTED

Up to this point, we have described typical graph searching methods. However, since the problem has a state space representation, searching the state space can often be made more efficient by using information (or knowledge) regarding the problem. For example, in the maze problem, if we know the position of the goal relative to the starting point, we might select paths which lead as far as possible in that direction. Also, in the 8-puzzle, we might look at the positions of pieces which are different from the goal state, and try to make them approach the goal state. In this way, if we can get to know which states are close to the goal, we can use these to direct the search.

If we can predict the cost to the goal, let us consider how we can use this information. It is important to note that this prediction need not necessarily be correct. It might happen, in the case of a maze, that a path in the direction of the goal might turn out to be a dead-end, increasing the searching time. However, in many cases prediction of the cost is valuable in searching. Such knowledge, which has no guarantee of being completely correct, but which is useful in the great majority of cases, is called *heuristic* knowledge. A search using heuristic knowledge is called a heuristic search.

We shall give three typical examples of heuristic searches. We shall write the inferred value of the cost from a node n of the graph, corresponding to a given state, to the goal as $\hat{h}(n)$. We assume that $\hat{h}(n)$ does not alter during the progress of the search.

3.4.1 Hill-climbing method

In searching a graph, the strategy of trying to reach the goal by choosing those nodes which are predicted to be nearest to the goal is called *hill-climbing*. Specifically, in order to choose the next node after n, we calculate $\hat{h}(n_i)$ for the child nodes $(n_1, n_2, \ldots, n_m)$ of n, and take the node for which $\hat{h}(n_i)$ is the least, as the next node. A hill-climbing search algorithm is as follows:

Procedure *hill-climbing*

1. $n :=$ start node.
2. *LOOP*: **if** *goal* (n) **then** *exit* (*success*).
3. Expand n, compute $\hat{h}(n_i)$ for all child nodes n_i and take the child node which gives the minimum value as *nextn*.
4. **If** $\hat{h}(n) < \hat{h}(nextn)$ **then** *exit* (*fail*).
5. $n := nextn$.
6. **Goto** *LOOP*.

If we imagine the goal as the summit of the hill, and $\hat{h}(n)$ as the difference in heights between n and the summit, the hill-climbing procedure corresponds to climbing the hill by always going upwards. If there is only one hill, this method must reach the goal. However, if there are many smaller hills other than the goal, the procedure may go up to the top of one of these hills and not advance any further.

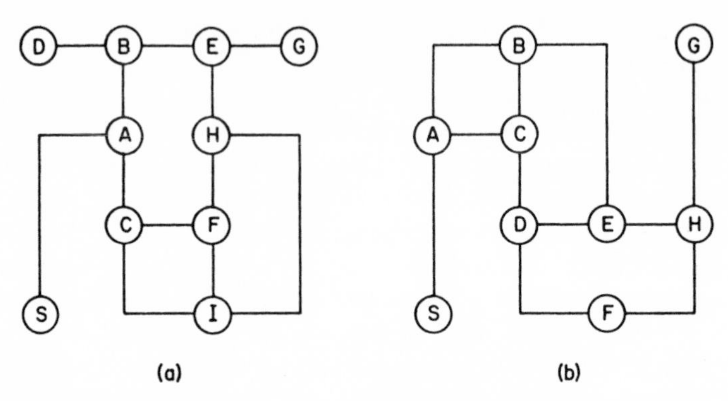

Fig. 15. Example of search using hill-climbing method. S is the starting point and G is the goal. The method succeeds in (a) and fails in (b)

An example of a problem where the hill-climbing method succeeds, and an example of a problem where it fails, are shown in Fig. 15. (Fig. 15(a) is the same as Fig. 2.). In both cases the cost from any location n to the goal is inferred using the following formula:

$$\hat{h}(n) = |x \text{ coordinate of goal} - x \text{ coordinate of } n| + |y \text{ coordinate of goal} - y \text{ coordinate of } n|$$

Suppose we now perform a search using the hill-climbing procedure on a graph in which the positions corresponding to stages of these problems are represented by nodes on a graph. In the case of Fig. 15(a), the goal node can be reached by following the nodes in order. The node sequence is S → A → B → E → G. This path is the optimum solution. Obviously the hill-climbing procedure restricts the range of the search. In the case of Fig. 15(b), however, the search fails after proceeding as follows: S → A → C → B; now B corresponds to the summit of a hill which is not the goal. In order to reach G, we have to go to the point E, which is further away from G (descending the hill).

Thus, if there are peaks other than the goal the hill-climbing method may fail. However, it may succeed. If in Fig. 15(b) the starting point is F, G will be reached.

3.4.2 Best-first search

In the hill-climbing method, if there are peaks (minimum values of $\hat{h}(n)$) other than the goal, the search cannot proceed any further after reaching these. To overcome this drawback, a technique known as the *best-first search*, in which searching is carried out after making an overall examination of the state space, has been developed.

The hill-climbing procedure corresponds to walking around the graph on one's own and choosing, from one's immediate environs, the node which one thinks is closest to the goal.

The best-first search procedure takes all the nodes that have been examined up to that time point and expands the node which is nearest to the goal. This is in contrast to the optimal-search procedure, which was described in the previous section, which expands the node nearest to the goal of the nodes that have been obtained so far. In other words, the best-first search procedure aims to minimize the labour which will be necessary in the future, irrespective of the labour which was spent on searching in the past.

The best-first search algorithm is very similar to the optimal-search algorithm. However, whereas in the latter the cost $\hat{g}(n)$ of the path to a node n may change during the progress of the search, in the case of the former $\hat{h}(n)$ is constant. Thus it is not necessary to change the pointers. A best-first search algorithm for searching a graph is given below:

Procedure best-first search
1. Put starting node S into *open*.
2. *LOOP*: **if** *open* = empty **then** *exit* (*fail*).
3. n := *first*(*open*).
4. **If** *goal*(n) **then** *exit* (*success*).
5. *Remove*(n, *open*).
   Add(n, *closed*).
6. Expand n, and generate all the child nodes. Of the child nodes, only put those into *open* that are not contained in *open* or *closed*. Give each of these a pointer to n. List the nodes of *open* in order of smallest $\hat{h}(n)$.
7. **Goto** *LOOP*.

If this procedure is applied to Fig. 15(a), the search which is carried out is exactly the same as with the hill-climbing method. The change in the *open* list when this procedure is applied to Fig. 15(b) is as follows (the figures in brackets show the value of $\hat{h}(n)$):

$$(S(6)) \rightarrow (A(4)) \rightarrow (B(2)\ C(3)) \rightarrow (C(3)\ E(3)) \rightarrow (E(3)\ D(4)) \rightarrow (H(2)\ D(4)) \rightarrow (G(0)\ D(4)\ F(4))$$

To perform the best-first search as efficiently as possible, we have to make the inference of the cost to the goal as accurate as possible. That is, we have to make an inference that is as close as possible to the true cost, using knowledge about the problem. We give a simple example of this below.

Example: Missionaries and cannibals problem

The state of the problem, which we discussed in Section 2.4, is expressed by three quantities: the number of missionaries on the left bank, the number of cannibals on the left bank, and the position of the boat. These are written as (M_L, C_L, p). As already mentioned, the problem is expressed by $(N, N, 1) \rightarrow (0, 0, 0)$. The state space has already been simplified, and is restricted to $(M_L = 0) \vee (M_L = N) \vee (M_L = C_L)$. Consider the case where $N = 5$ and k (the crew of the boat) is 3. The state space of this problem is shown in Fig. 16(a). If all combinations of M_L, C_L and p were allowed, $5 \times 5 \times 2$ (= 50) nodes would be necessary, but the simplification reduces this to 24.

Initially, let us take $\hat{h}(n) = M_L + C_L + p$ as the inferred cost of reaching the goal (0, 0, 0) from the state (M_L, C_L, p) of a node n. Fig. 16 shows the nodes generated by the search. The path obtained is shown by the heavy line in Fig. 16(b). Fig. 16(a) shows all the nodes that are generated, but contains some nodes that are not expanded. We can see that some searching time has been saved compared with searching all paths.

Now let us change the definition of $\hat{h}(n)$. We might say that if some people are left on the left bank, the state where the boat is at the left bank is nearer the

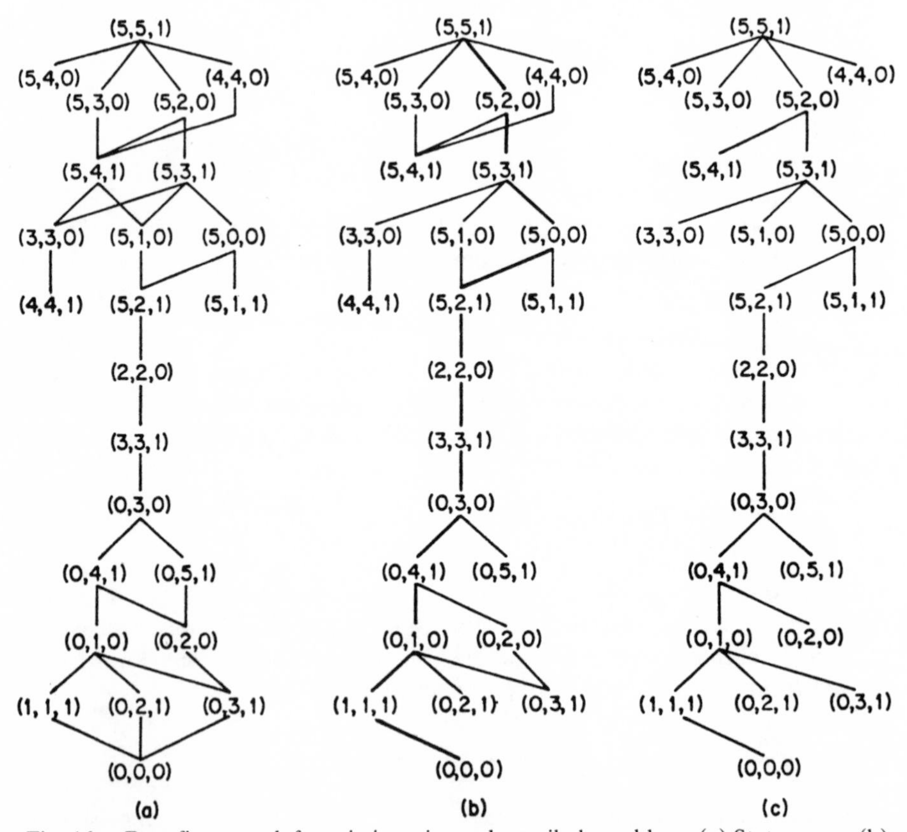

Fig. 16. Best-first search for missionaries and cannibals problem. (a) State space, (b) $\hat{h}(n) = M_L + C_L + p$, and (c) $\hat{h}(n) = M_L + C_L - 2p$

goal than the state where the boat is at the right bank. So we take $\hat{h}(n) = M_L + C_L - 2p$. The result of the search is shown in Fig. 16(c). One fewer nodes are generated than in Fig. 16(a), but there is an appreciable reduction in the number of nodes that are expanded.

Consider now the change in $\hat{h}(n)$ on moving from one node to the next. Since the boat holds three people, in moving from the left bank to the right bank, the most $\hat{h}(n)$ can change by is 1. One strategy would therefore be to expand the node n, and, if a child node n for which $\hat{h}(n_i) = \hat{h}(n) - 1$ is obtained, to take this as the next node. (If this resulted in a dead-end, we would have to back up and select another child node.) If we adopt this strategy, we can make a further reduction in the number of points generated, greatly improving the efficiency of the search.

3.4.3 Optimum solution when cost can be predicted

In Section 3.3 we described a method of finding the optimum solution without using knowledge about the problem. We shall now describe a method of finding the optimum solution when the cost to the goal can be inferred.

Take an arbitrary node n on the graph, and let $g(n)$ be the cost of the optimum path from the starting node S to n. We call the cost of the optimum path from n to the goal node $h(n)$. If there is no path, $g(n)$ or $h(n)$ is taken as infinitely large. The cost $f(n)$ of the optimum path through n is given by the following:

$$f(n) = g(n) + h(n) \tag{4}$$

and the problem is to find the optimum path (the path that gives $f(\mathrm{S})$) from the starting node. If $f(n)$ is known exactly, the solution can be obtained by following the nodes of least $f(n)$ from S. In practice, neither $g(n)$ nor $h(n)$ is accurately known, as a search is required.

As in the previous examples, the value of the inferred cost $\hat{h}(n)$ to the goal is given for all nodes. We take the inferred value $\hat{g}(n)$ of $g(n)$ for the nodes that have already been generated as the minimum cost, of the paths to n that are known so far. The inferred value $\hat{f}(n)$ of the cost of the optimum path going through the node n is given by the following:

$$\hat{f}(n) = \hat{g}(n) + \hat{h}(n) \tag{5}$$

The strategy of searching the graph using equation (5) as an evaluation function is termed the *A-algorithm*. It may be written as follows:

Procedure *A-algorithm*
1. Enter the starting node in *open*. $\hat{f}(\mathrm{S}) := \hat{h}(\mathrm{S})$.
2. *LOOP*: **if** *open* = empty **then** *exit* (fail).
3. n := *first*(*open*).
4. **If** *goal*(n) **then** *exit*(success).

5. *Remove*(n, *open*).
   Add(n, *closed*).
6. Expand n. For all child nodes n, calculate $\hat{f}(n, n_i) = \hat{g}(n, n_i) + \hat{h}(n_i)$ using the cost $\hat{g}(n, n_i)$ through n to n_i from S. Put nodes that are contained in *open* or *closed* into *open*, and set pointers to n. For nodes contained in *open*, compare $\hat{f}(n_i)$ and $\hat{f}(n, n_i)$ before expanding n. If $\hat{f}(n, n_i)$ is smaller, make $\hat{f}(n_i) = f(n, n_i)$, and set pointer from n_i to n. If the child node n_i is contained in *closed*, and $\hat{f}(n, n_i) < \hat{f}(n_i)$, make $\hat{f}(n_i) = \hat{f}(n, n_i)$, set a pointer to n from n_i, and put n_i in *open*. Lastly, list the nodes in *open* in order of least $\hat{f}$.
7. **Goto** *LOOP*.

Note: if we put $\hat{h}(n) \equiv 0$, the A-algorithm exhibits the same behaviour as the optimal-search algorithm described in Section 3.3 (nodes in *closed* never get put back onto *open* again).

If the graph is finite it can be shown, by the same method as for the proof of Theorem 1, that the A-algorithm will inevitably find a path to the goal node, if a path from the starting node to the goal node exists.

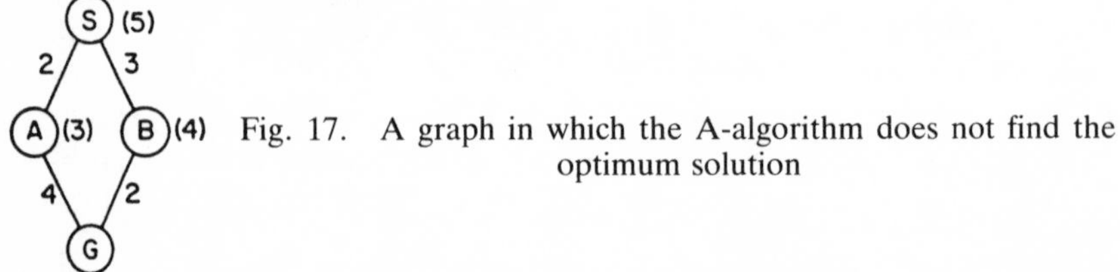

Fig. 17. A graph in which the A-algorithm does not find the optimum solution

However, there is no guarantee that the A-algorithm will obtain the optimum solution. Fig. 17 shows a simple example of this. The figures in the brackets adjacent to the nodes are the $\hat{h}(n)$ for the node n. First we expand S to obtain child nodes A and B. We compute the respective evaluation functions using equation (5). Since $\hat{f}(A) = 2 + 3$, and $\hat{f}(B) = 3 + 4$, the *open* list is (A B). We then fetch A from *open*, and expand it, obtaining G. $\hat{f}(G) = 6 + 0$, so the *open* list becomes (G B). When we now fetch G, the search ends. The solution is the path S → A → G. Obviously this is not the path of minimum cost. The reason why the minimum cost path S → B → G was not found is that $f(B)$ was greater than $\hat{f}(B)$, so it did not come at the head of *open*.

We can prove that, if $\hat{h}(n)$ exceeds $h(n)$, the A-algorithm will find the optimum solution. (This proof is given later.) If the inferred cost is the lower bound of the true cost ($\hat{h}(n) \leqslant h(n)$), we call the A-algorithm using this the A*-algorithm.

An example of searching of a graph using the A*-algorithm will now be given. The graph to be searched is shown in Fig. 18. As in Fig. 17 the value of $\hat{h}(n)$ is given in brackets adjacent to each node. The changes in the *open* list obtained by using the A-algorithm procedure are as follows ($\hat{f}$ for each node is

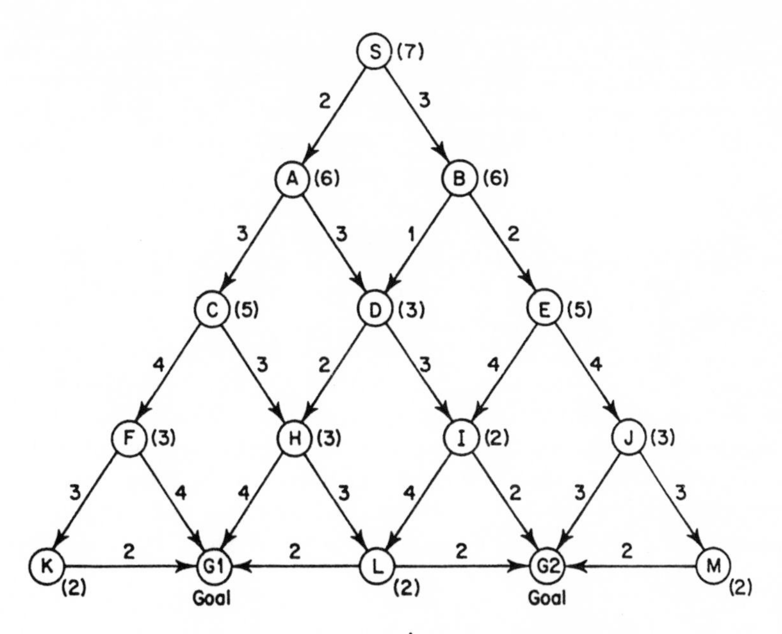

Fig. 18. Graph for which $\hat{h}(n) \leqslant h(n)$ is satisfied

shown in brackets):

(S(7)) → (A(8) B(9)) → (D(8) B(9) C(10)) → (B(9) C(10) H(10) I(10)) → (D(7) C(10) E(10) H(10) I(10)) → (H(9) I(9) C(10) E(10)) → (I(9) G1(10) C(10) E(10) L(11)) → (G2(9) G1(10) C(10) E(10) L(11))

This procedure terminates when G2 is fetched from the *open* list.

The solution S → B → D → I → G2 is indeed the optimum solution ($h(S) = 9$). It should be noted that a node n that satisfies $\hat{h}(n) \leqslant h(S)$ must exist in the *open* list and for at least the nodes on the optimum path $\hat{h}(n) \leqslant h(S)$. This inequality will be proved later. It should also be noted that, although during the progress of the search node D gets entered into the *closed* list once, subsequently the computer returns to D and finds a path with a lower cost, and so again returns it to the *open* list.

Since $\hat{h} \equiv 0$ is surely a lower bound of h, the optimal-search procedure using this is a special case of the A*-algorithm. The A*-algorithm described below therefore corresponds to an optimal-search procedure not using heuristic knowledge.

3.4.4 Characteristics of the A*-algorithm

Since the A*-algorithm is a special case of the A-algorithm, it must have all the characteristics of the latter. Specifically, if the graph is finite and there is a path from the starting node to the goal, it must terminate by finding a path to the goal node. We shall now show that, even if the graph is infinite, the A*-algorithm finds the optimum solution.

Lemma 1.

Representing the starting node by S, before the A*-algorithm terminates, there will always be a node n that satisfies $\hat{f}(n) \leqslant f(\mathrm{S})$ in the *open* list.

Proof. The optimum path from S to the goal node is represented by a series $(n_0, n_1, \ldots, n_m)$ of nodes ($n_0 = \mathrm{S}$, $n_m =$ goal node). Initially S is put in *open*, and the algorithm is not yet terminated, so there is a node on the optimum path in the *open* list. Of the nodes in the *open* list, let us use n to represent the node that appears first in the sequence of nodes constituting the optimum path. The ancestor nodes of this node already appear in *closed*, so the optimum path as far as the parent node of n has already been found. Since n is on the optimum path to the goal node, the optimum path up to n has also been found. Therefore:

$$\begin{aligned}\hat{f}(n) &= \hat{g}(n) + \hat{h}(n) \\ &= g(n) + \hat{h}(n)\end{aligned}$$

Since, in the A*-algorithm, $\hat{h}(n) \leqslant h(n)$:

$$\hat{f}(n) \leqslant g(n) + h(n) = f(n)$$

n is on the optimum path, so $\hat{f}(n) \leqslant f(\mathrm{S})$. That is, there exists an n that satisfies $f(n) = f(\mathrm{S})$

Next, we prove that the A*-algorithm terminates.

Lemma 2.

If the graph has a path from the starting node to the goal node, a path to the goal node can always be found by the A*-algorithm.

Proof. We have already stated that, if the A*-algorithm terminates, it will always find a path to the goal node. We should therefore now show that it must always terminate. Suppose the A*-algorithm does not terminate. Let us represent the length of an edge by 1, and the minimum value of the length of the path from the starting node to node n by $d(n)$. As the number of branches from a single node is finite, as the search proceeds $d(n)$ for any node n in the *open* list becomes larger than an arbitrary number M. Since the cost of the edge is positive, if we take its minimum value as ε, we have $g(n) \geqslant d(n)\varepsilon$.

Since $\hat{g}(n) \geqslant g(n)$, and $\hat{f}(n) \geqslant \hat{g}(n)$, $\hat{f}(n) \geqslant d(n)\varepsilon$. If we take $M = f(\mathrm{S})/\varepsilon$, as the search proceeds $d(n)$ becomes greater than M, so we have $\hat{f}(n) \geqslant f(\mathrm{S})d(n)/M > f(\mathrm{S})$. This contradicts Lemma 1. In other words, the A*-algorithm terminates.

Theorem 3

If the graph has a path from the start node S to the goal node, the A*-algorithm will always terminate by finding the optimum solution.

Proof. We know from Lemma 2 that the A*-algorithm terminates. We have also stated that, if it terminates, it always finds a path to the goal node. We therefore have to prove that, when the A*-algorithm terminates by finding the goal node G, that is the optimum solution. If we assume that it is not the optimum solution, $\hat{f}(G) > f(S)$.

From Lemma 1, immediately before the search terminates there exists a node n in the *open* list that satisfies $\hat{f}(n) \leqslant f(S)$. Since, for such a node n, $\hat{f}(n) < \hat{f}(G)$ at that time point, the A*-algorithm would select n rather than G. This contradicts the assumption that the search would terminate by choosing G.

Corollary 1

$\hat{f}(n')$ for node n' expanded by the A*-algorithm is larger than the cost of the optimum solution.

Proof. Since n' was expanded, it is not the goal node, and the search has not terminated. From Lemma 1 the *open* list contains a node n that satisfies $\hat{f}(n) \leqslant f(S)$. The A*-algorithm fetches the node for which $\hat{f}$ is a minimum from the open list and expands it. That is, $\hat{f}(n') \leqslant \hat{f}(n)$. Therefore $\hat{f}(n') \leqslant f(S)$.

From Theorem 3 we can see that when $\hat{h}(n) \equiv 0$, the optimal search algorithm always finds the optimum solution.

We now examine the relationship between the heuristic function $\hat{f}(n)$ and the A*-algorithm. If $\hat{f}(n) = f(n)$, the solution can be found without expanding unnecessary nodes. It may be expected that, the nearer $\hat{f}(n)$ is to $f(n)$, the less will be the number of nodes to be expanded. Let us suppose that we have A*-algorithms using different evaluation functions A_1 and A_2 respectively. Let us assume that A_1 uses $\hat{f}_1(n)$, while A_2 uses $\hat{f}_2(n)$. If for all nodes n, $\hat{h}_1(n) > \hat{h}_2(n)$, we say that the knowledge of A_1 is greater than the knowledge of A_2.

Theorem 4

If there are two A*-algorithms A_1 and A_2, and if the knowledge of A_1 is greater than that of A_2, the nodes that are expanded by A_1 will always be expanded by A_2.

Proof. We use the method of mathematical induction. The theorem holds for the end-point of a path of length 0 from the starting node. (If S is the goal node, neither A_1 nor A_2 is expanded, otherwise both are expanded.) Next we show that if the theorem holds for the end-points of all paths of length k from S, it also holds for the end points of paths of length $k + 1$. Suppose that there is a node which is expanded by A_1 but not expanded by A_2. Call this node n. Since A_2 expands the parent node of n, when A_2 terminates, n is left in the *open* list. Since n is not selected, when the algorithm terminates, the following relationship holds:

$$\hat{f}_2(n) \geqslant f(S)$$

Since $\hat{f}_2(n) = \hat{g}_2(n) + \hat{h}_2(n)$:

$$\hat{h}_2(n) \geqslant f(S) - \hat{g}_2(n)$$

On the other hand, since A_1 has expanded n, at that time-point:

$$\hat{f}_1(n) \leqslant f(S)$$

Since $\hat{f}_1(n) = \hat{g}_1(n) + \hat{h}_1(n)$:

$$\hat{h}_1(n) \leqslant f(S) - \hat{g}_1(n)$$

Since any node that is expanded by A_1 and which is a node on the path of length k will also be expanded by A_2:

$$\hat{g}_2(n) \leqslant \hat{g}_1(n)$$

that is:

$$\hat{h}_1(n) \leqslant f(S) - \hat{g}_1(n) \leqslant f(S) - \hat{g}_2(n) \leqslant \hat{h}_2(n)$$

This contradicts the assumption that $\hat{h}_1 > \hat{h}_2$. Therefore the assumption that n exists is not true.

From Theorem 4 we can see that the A*-algorithm using heuristic knowledge $h(n)$ expands less nodes than a search for the optimum solution not using $\hat{h}$ ($\hat{h} \equiv 0$). Fig. 19 shows a marked example of this. In both cases, the optimum path (S C J G7) and its cost 8 is obtained but, when $\hat{h} \equiv 0$, all the nodes other than the goal nodes G1–G8 are expanded. This corresponds to finding the minimum value of the cost after finding all the paths. In contrast, if the values given in the brackets in the figure are used for $\hat{h}$, no nodes are expanded other than those lying on the optimum path. However, things do not necessarily work out so well in practice.

3.4.5 Improvements to the A*-algorithm

We have seen that use of the A*-algorithm using a heuristic function $\hat{h}(n)$ enables fewer points to be expanded than would be expanded with a simple search for the optimal solution with $h \equiv 0$. The number of nodes expanded is one criterion of searching efficiency. However, as shown by step 6 of the A-algorithm it can occur that nodes which have already been expanded once

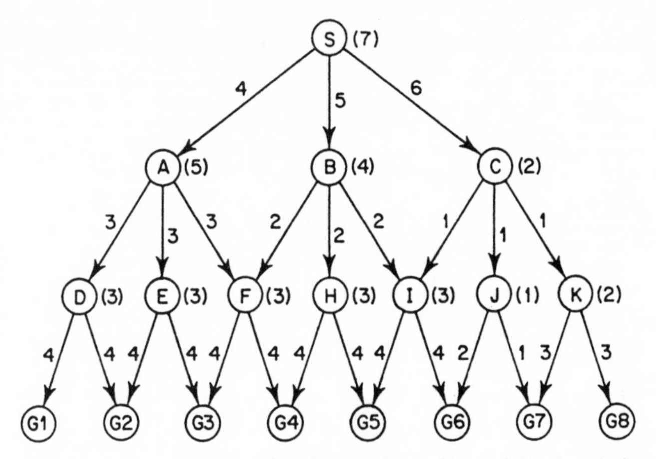

Fig. 19. Example showing the marked effect of the heuristic function $\hat{h}$

and put into the *closed* list may again be returned to the *open* list. In other words, it is possible for the same node to be expanded more than one time. Even if the number of nodes expanded is reduced, if the number of times they are expanded is increased, the searching efficiency will be poor.

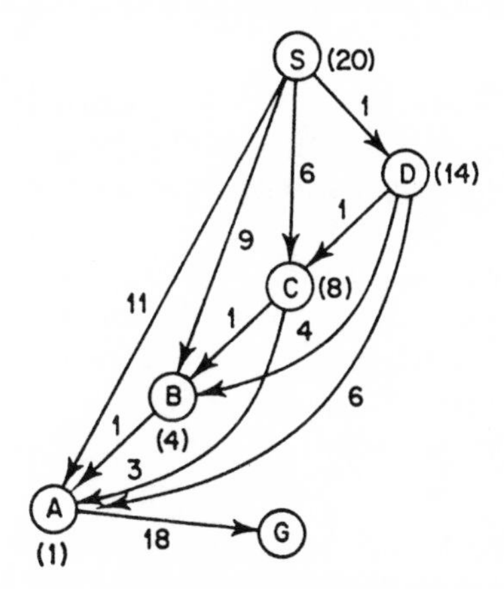

Fig. 20. Graph in which a node is expanded several times by the A*-algorithm

Fig. 20 shows an example of a graph in which the A*-algorithm expands the same node several times. The operation of this algorithm is shown in Table 2. Each row of the table shows the nodes in the *open* and *closed* lists and their inferred costs $\hat{f}$. If the node is in the *closed* list, this is shown by the symbol†. Otherwise the node is in the *open* list. Initially, only the starting node S is in the *open* list, so $\hat{f}$(S) is given in row 1 of the table. Next, S is put into *closed*, and

TABLE 2
Operation of the A*-algorithm for Fig. 20

S	A	B	C	D	G
20					
20†	12	13	14	15	
20†	12†	13	14	15	29
20†	11	13†	14	15	29
20†	11†	13†	14	15	28
20†	10	11	14†	15	28
20†	10†	11	14†	15	27
20†	9	11†	14†	15	27
20†	9†	11†	14†	15	26
20†	8	9	10	15†	26
20†	8†	9	10	15†	25
20†	7	9†	10	15†	25
20†	7†	9†	10	15†	24
20†	6	7	10†	15†	24
20†	6†	7	10†	15†	23
20†	5	7†	10†	15†	23
20†	5†	7†	10†	15†	22
20†	5†	7†	10†	15†	22†

the nodes A, B, C, and D which are obtained by expanding S are put into *open*. Next, the node A for which $\hat{f}$ is a minimum is fetched from the *open* list and put in the *closed* list, and A is expanded. When this procedure is repeated, the node is expanded a total of 17 times. If the number of nodes is N, it is known that, in the worst case, the number of times the nodes are expanded is $O(2^N)$.‡

In contrast, if we do not use the heuristic knowledge, all the nodes are expanded once only. Although the same number of nodes are expanded by the A*-algorithm, they are expanded many times.

In order for the A*-algorithm not to expand the same node twice, we may set up the following condition:

$$\hat{h}(n_i) \leqslant \hat{h}(n_j) + c(n_i, n_j)$$

for any node n_i of the graph and its child node n_j. This condition is called the *monotone restriction*. This fact is expressed by the following theorem.

Theorem 5

If the monotone restriction is satisfied, at the time point when the node n is expanded by A*, the optimum path up to n has already been found.

‡Martelli, A., 'On the complexity of the admissible search algorithm', *Artificial Intelligence*, **8**, 1977, 1–13.

Proof. Represent the optimum path up to n by the node series $P = (n_0, n_1, \ldots, n_k)$ (where $n_0 = S$, and $n_k = n$). At the time point when a node n other than the start node is fetched from the *open* list to be expanded, let us assume that P has not been found. At this time point, of the nodes in the *closed* list, there must be a node included in P (at least S is in the *closed* list). Of such nodes, let us take n_l as being the one which comes latest in the series P. Since n_{l+1} is in the open list, $n_{l+1} \neq n$. From the monotone restriction, for an arbitrary i:

$$g(n_i) + \hat{h}(n_i) \leqslant g(n_i) + \hat{h}(n_{i+1}) + c(n_i, n_{i+1})$$

Since n_i and n_{i+1} are on the optimum path:

$$g(n_{i+1}) = g(n_i) + c(n_i, n_{i+1})$$

and so:

$$g(n_i) + \hat{h}(n_i) \leqslant g(n_{i+1}) + \hat{h}(n_{i+1})$$

Since this inequality holds for all neighbouring nodes of P, if we use the inequality from $i = l$ to $k - 1$, we have:

$$g(n_{l+1}) + \hat{h}(n_{l+1}) \leqslant g(n_k) + \hat{h}(n_k)$$

that is:

$$g(n_{l+1}) + \hat{h}(n_{l+1}) \leqslant g(n) + \hat{h}(n)$$

On the other hand, since A* has fetched n from the open list:

$$\hat{g}(n) + \hat{h}(n) \leqslant \hat{f}(n_{l+1})(= g(n_{l+1}) + \hat{h}(n_{l+1}))$$

From the above two inequalities, $\hat{g}(n) \leqslant g(n)$. In other words, the optimum path to n has been found.

Since Fig. 20 does not satisfy the monotone restriction, sometimes the optimum path up to the expanded node will not have been found, so this node will again be placed in the *open* list. Theorem 4 is obtained if $\hat{h} \equiv 0$ because the monotone restriction is then satisfied. Clearly Theorem 2 is only a special case of Theorem 5.

An improved algorithm has been proposed aimed at finding the solution by expanding the nodes fewer times than in the case of the A*-algorithm. This improved algorithm is based on the following fact. From Lemma 2, before the search terminates, a node n satisfying $\hat{f}(n) \leqslant f(S)$ is present in the *open* list. Assuming that the A*-algorithm has expanded the node n', we have $\hat{f}(n') \leqslant \hat{f}(n) \leqslant f(S)$. If, of the nodes expanded up to a given time point by the search, the node of maximum inferred cost immediately before expansion is m, $\hat{f}(m) \leqslant f(S)$. If $\hat{f}(n_i) < \hat{f}(m)$ for a node n_i in the *open* list at that time point, n_i will inevitably be expanded at some time. The reason for this is that if $\hat{f}(n_i) < \hat{f}(m)$, we also have $\hat{f}(n_i) < f(S)$, so the goal node G ($\hat{f}(G) = f(S)$)

cannot be fetched from the *open* list while $\hat{f}(n_i)$ is still left open. It will therefore be worth while to expand all the nodes that satisfy $\hat{f}(n_i) < \hat{f}(m)$ after m has been expanded. If there are several such nodes, we can prevent the same node being expanded twice or more by arranging the nodes during the expansion in order of least $\hat{g}$ instead of using $\hat{f}$. This searching procedure is called the *modified* A*-algorithm. As shown below, the difference between this procedure and the A-algorithm lies in steps 1–3:

Procedure modified A-algorithm
1. Put start node in open list. $\hat{g}(S) := 0, \hat{f}(S) = \hat{h}(S), f_m := 0.$
2. *LOOP*: **if** *open* = empty **then** *exit*(*fail*).
3. *nset* := $\{n_i | \hat{f}(n_i) < f_m\}$; make the set of nodes in the open list that satisfies $\hat{f} < f_m$ *nset*.
   If *nset* $\neq$ empty, **then** call the node of least $\hat{g}$ which is an element of *nset n*.
   else $n := first\,(open), f_m := \hat{f}(n)$.
4. The rest is the same as the A-algorithm.

TABLE 3
Operation of the modified A*-algorithm on the graph of Fig. 20 (values of $\hat{g} + \hat{h}$ shown from S to G)

S	A	B	C	D	G	f_m
0 + 20						0
0 + 20†	11 + 1	9 + 4	6 + 8	1 + 14		20
0 + 20	7 + 1	5 + 4	2 + 8	1 + 14†		20
0 + 20	5 + 1	3 + 4	2 + 8†	1 + 14†		20
0 + 20	4 + 1	3 + 4†	2 + 8†	1 + 14†		20
0 + 20	4 + 1†	3 + 4†	2 + 8†	1 + 14†	22 + 0	20
0 + 20	4 + 1†	3 + 4†	2 + 8†	1 + 14†	22 + 0†	22

The nodes in the *open* and *closed* lists and values of $\hat{g}$, $\hat{h}$, and f_m when Fig. 20 is searched using this procedure are shown in Table 3. The dagger symbol (†) indicates that the node is in the *closed* list. It can be seen that there is a considerable reduction in the number of times the nodes are expanded compared with the case of the A*-algorithm (Table 2).

In general, the number of times the nodes are expanded using the modified A*-algorithm is either less than or equal to that using the A*-algorithm. It can be proved that in the worst case the number of times is $O(N^2)$.‡

3.5 USE OF CONSTRAINTS

In Section 2.4 we gave an example of simplification of problem representation using constraints. We shall now describe a method whereby

‡ Martelli, A., *loc. cit.*

constraints may be used to restrict the range of searching during the progress of a search.

In voice recognition or image recognition, the whole is recognized by extracting suitable characteristics from the input pattern and interpreting these various characteristics. For example, one technique is to determine the phonemes by dividing the voice waveform into several portions, finding possible candidates for the phonemes corresponding to each element of this waveform, and determining the phonemes for these respective elements in such a way that the sequence of these elements forms a non-contradictory utterance. The constraint operating between the elements is that the whole waveform should constitute a non-contradictory utterance. That is to say, the various elements have to form words and the sequence of words has to make a grammatically correct utterance.

Taking the various elements as $n_1, \ldots, n_m$, assume that there are b_i possible interpretations (e.g. phonemes) for an element n_i. If we had to generate all possible interpretations and examine whether they satisfied the constraints, there would be $b_1 \times b_2 \times \ldots \times b_m$ cases to examine.

If we represent the problem by a tree, taking the start node as n_1, the b_1 different interpretations of n_1 correspond to b_1 edges originating from the node n_1. The tree is constructed by drawing edges corresponding to the n_2 interpretations from the end-points of these edges and repeating this process for each node. The lowest node (end-point) of the tree corresponds to an utterance. This method is very inefficient if all possible utterances are generated. The intermediate nodes of the tree correspond to some elements of the utterance having been elucidated. If the sequence of elements that have been elucidated does not correspond to a word, further searching is pointless.

Restricting the range of a search while it is being carried out, by using constraints, is an important strategy in solving problems in artificial intelligence.

We shall formalize the method of using constraints by applying it to the problem of labelling nodes on a graph.

3.5.1 Problem of labelling nodes on a graph

Consider the problem of interpreting an object. Let us assume that the object can be analysed into several elements $n_1, n_2, \ldots, n_m$, and that the object can be interpreted by interpreting these various elements. For each element n_i, a set L_i of possible interpretations is given. Whether or not an interpretation $l_i \in L_i$ of n_i and an interpretation $l_j \in L_j$ of n_j are possible, or are mutually contradictory, is determined by the fixed relationship b_{ij} that exists between two elements n_i and n_j. If the interpretation is possible, we write this fact as $(l_i, l_j) \in D(b_{ij})$.

If we represent the various elements n_i by nodes, and the relationships b_{ij} by edges joining nodes n_i and n_j, the relationship between all the elements can be expressed by a graph. If b_{ij} does not apply any constraint on the interpretation of n_i and n_j, we take it that there is no edge between n_i and n_j. We call the giving of a suitable interpretation l_i from L_i to the node n_i 'assigning the *label* l_i to the node n_i'. The problem is to give all the nodes of the graph a label, and to ensure that, for all the edges b_{ij}, the labels l_i and l_j at the end-points of these edges satisfy $(l_i, l_j) \in D(b_{ij})$.

Fig. 21 shows a simple example. The set of possible labels is shown in brackets at each node, and the constraint which the labels of the nodes at both ends of the edge have to satisfy is shown along the edge.

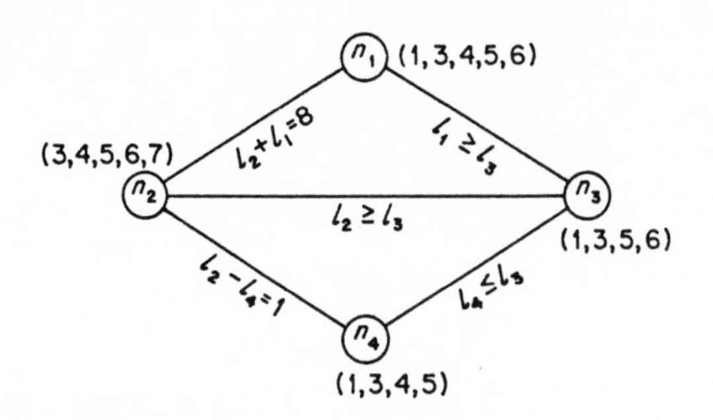

Fig. 21. Problem of labelling the nodes of a graph. Possible labels are shown in brackets

3.5.2 Searching using propagation of constraints

A method of restricting the range of searching during progress of the search using the constraints will now be explained using the example of Fig. 21. The basic strategy is to narrow down the set of possible labels of the nodes using the constraints corresponding to the edges. First we attach the set L_i of candidate labels to a suitable node n_i (in this example, to n_1). We write this as L_i^0 to show that it is the initial set of labels for n_i. In this example $L_i^0 = (1, 3, 4, 5, 6)$. Next we look at the nodes that are connected to n_i by edges (adjacent nodes), choose a suitable node n_j (n_2 in this example), and remove from L_j all the labels which do not correspond to any element of L_i^0. This produces the new set of labels L_j^1. That is:

$$L_j^1 = \{l_j | l_j \in L_j \wedge \exists\, l_i[l_i \in L_i \wedge (l_i, l_j) \in D(b_{ij})]\}$$

In this example, in the case of the candidate label 6 of n_2, a label 2 which would satisfy the constraint $l_2 + l_1 = 8$ does not appear in L_1^0; so the candidate label 6 is removed, and we have $L_2^1 = (3, 4, 5, 7)$. When L_j^1 has been determined, we check to see whether the sets of labels of adjacent nodes contain any that are inconsistent with L_j^1. In this example, we were unable to find another label in L_2^1 satisfying the constraint and corresponding to the label 6 of L_1^0; so the label 6 was removed. The new set of labels for n_1 is now $L_1^1 = (1, 3, 4, 5)$. As shown in Fig. 22(a), the constraint b_{ij} is applied to the

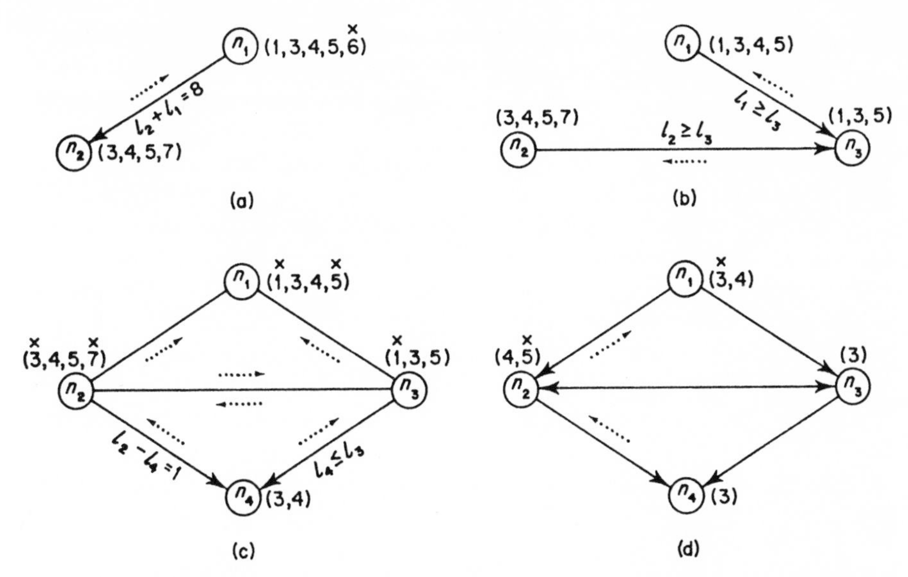

Fig. 22. Propagation of constraints. Arrows show the initial directions of propagation, and dotted arrows show subsequent propagation. Crosses indicate that a label is removed from the set of possible labels

set of labels L_1^0 of n_1 to determine the set of labels L_2^1 of n_2; then b_{ij} is again used to determine, from L_2^1, the set of labels L_1^1 of n_1. That is, the constraint is propagated from n_1 to n_2 and then from n_2 to n_1, narrowing down the number of possible labels in the process.

Incidentally, it should be noted that there is no possibility of the constraint being again propagated from n_1 to n_2. The reason is that conformity has now been established between the elements of L_2^1 and L_1^0. Since any element in L_1^0 that did not match an element in L_2^1 was removed, there is no need to remove any element from L_2^1. Next, we choose a suitable node adjacent to n_j. Preferably, this should be a node that is adjacent to as many as possible of the nodes that have already been searched. In this example, from the nodes adjacent to n_2 we select n_3, since it is also adjacent to n_1. We propagate the constraint in the same way as described above from the adjacent nodes (in this case n_1 and n_2) that have already been examined. As shown in Fig. 22(b), the element 6 of L_3^0 is removed by the propagation from n_1. No change is caused by the propagation from n_2. So $L_3^1 = (1, 3, 5)$.

Next, the constraints are propagated from n_3 to n_1 and n_2. However, in this case this results in no change, so $L_1^2 = L_1^1$, and $L_2^2 = L_2^1$.

Finally, we carry out the same process for n_4. As shown in Fig. 22(c), because of the propagation of the constraint from n_2, we have $L_4^3 = (3, 4)$. There is no effect from n_3. Propagation of the constraint from n_4 to n_2 and n_3 results in $L_2^4 = (4, 5)$ and $L_3^4 = (3, 5)$. The constraint between n_2 and n_3

does not alter L_2^4 and L_3^4. Next, the constraint is propagated from n_2 and n_3 to n_1, giving $L_1^4 = (3, 4)$. This result is again transmitted to n_2, n_3, and n_4 to give $L_3^5 = (3)$, and $L_4^5 = 3$ as shown in Fig. 22(d). The number of possible labels is again reduced by propagation of the constraint from n_4 to n_2 and n_1. Thus we find that labels (l_1, l_2, l_3, l_4) are (4, 4, 3, 3). It can be seen that no further change takes place.

In general, the labels will not be uniquely determined by such a search procedure. In this example, if $L_4 = (2, 3, 4, 5)$, the nodes n_1 to n_4 will be left with the respective sets of labels (4, 5), (3, 4), (3) and (2, 3) when the searching procedure is carried out using propagation of the constraints. The available possibilities cannot be reduced any further. When adjacent nodes each have several labels corresponding to them, the labels cannot be combined in an arbitrary fashion. In the above example, there are only two labelling possibilities, namely (4, 4, 3, 3) and (5, 3, 3, 2). Where there are a number of possible labels, as in this case, in order to decide which labelling schemes are possible, we have to select one of the labels and find whether there are suitable labels matching it at the other nodes.

3.5.3 Application to interpretation of a line drawing

A practical example of the problem of labelling nodes on a graph by propagation of constraints is the interpretation of a line drawing. Fig. 23 shows an example of a line drawing of a polyhedron (ignore the labels on the lines). On seeing this line drawing, we human beings are able to interpret it as a rectangular parallelepiped, with a cube with holes in, placed on top. Such interpretation can be represented by assigning labels to lines of the line drawing. The lines correspond to edges of the polyhedron. If only one of the pair of faces corresponding to the edge is visible, we indicate this by putting an arrow label on the edge, as shown in Fig. 23, adopting the convention that the visible face is to the right of the arrow. If both sides of the edge are visible, we

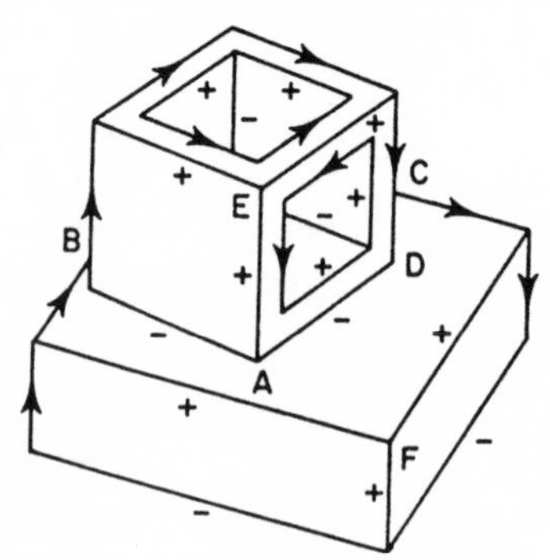

Fig. 23. Line labels in a line drawing

mark the edge with a '+' label if it is convex, and with a '−' label if it is concave. In Fig. 23 the rectangular parallelepiped is imagined to be placed on top of a table so its bottom lines have '−' labels. If the bottom of the parallelepiped were floating in space, they would have leftwardly pointing arrows.

The problem of interpretation of a line drawing is one of assigning labels to the drawing so that it represents a scene constructed of polyhedra in the real world. Though this problem seems to be simple, it is difficult to find a general algorithm for labelling a line drawing. In this chapter we shall describe a method of interpretation that will work if some restrictions are placed on the objects constituting the scene. The first restriction is that we always have three planes intersecting at the vertices of the polyhedron in question. If we consider Fig. 23 as made up of two polyhedra, four planes intersect at the vertex A. If the two polyhedra are stuck together at their faces so as to form a single polyhedron, three planes intersect at point A. Otherwise, this restriction is satisfied even if the upper cube is not considered as being in contact with the lower parallelepiped. The second restriction is that the line drawing is to represent the scene as viewed from a general position. For example, in Fig. 24 the edges AB and BC of the triangular prism appear as a straight line. If the viewing position is displaced a little horizontally, ABC is no longer a straight line. We exclude line drawings produced by viewing from such special positions.

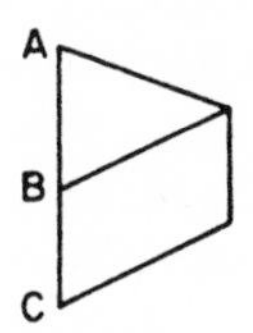

Fig. 24. Example of a non-general position

In a line drawing, the intersection of several lines is called a *junction*. The problem is to work out the way in which the labels are to be attached from the form of the junctions. For line drawings that satisfy constraints 1 and 2, the forms of the junctions and the labelling of the various lines are restricted to certain definite types.‡ Table 4 gives the names of the types of junction and their possible labels. We call the first labelling possibility of the L junction $L1$.

The problem of labelling a line drawing can be reduced to the problem of labelling the junctions shown in Table 4. The junctions in the line drawing correspond to the nodes of a graph, and the problem is reduced to one of labelling these nodes. The set of possible labels for each node is given by the form of the corresponding junction and is shown in Table 4. If two junctions are

‡ Huffman, D. A., 'Impossible objects as nonsense sentences', in Meltzer, B., and Michie, D. (eds.), *Machine Intelligence*, **6**, 1971, 295–323.

TABLE 4
Junction types and possible labels

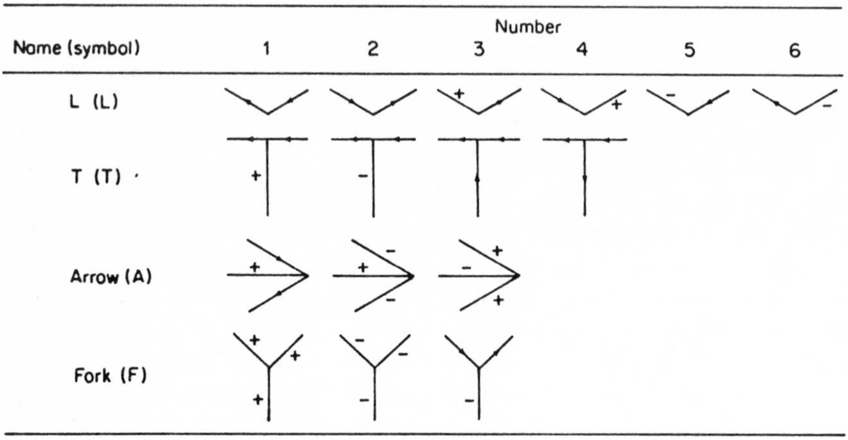

Name (symbol)	Number 1	2	3	4	5	6
L (L)						
T (T)						
Arrow (A)						
Fork (F)						

joined by a line, there is an edge between the corresponding nodes. The restriction on the edges is that the same labels should be attached to the lines corresponding to the edges. For example, if the labels A2 are attached to the node A in Fig. 23, the line AD has a '−' label. The labelling of the junction D at the other end of the line must therefore be L5. Thus the problem of interpretation of a line drawing is formalized as a problem of labelling the nodes of a graph.

First we assign possible sets of labels by choosing nodes (referred to below simply as junctions) corresponding to suitable junctions. Preferably, we choose junctions for which the set of labels has as few elements as possible. In the case of interpretation of a line drawning, the outside lines represent the boundaries of the polyhedron, so they will either have an arrow, meaning that the adjoining surface is on the right, on the other side of the shape shown in the drawing, or a '−' sign. The labelling possibilities for junctions at the ends of lines on the outside of the figure are therefore limited. Successive adjacent junctions are then searched, preferably choosing junctions for which the constraints are as strong as possible. There is a strong constraint across the horizontal stroke of the T junction in that the labels of both lines must be the same, but the vertical stroke can have any label, i.e. no constraint is propagated.

In Fig. 23, if we know that the polyhedron is resting on the table we immediately know the labels of the outside lines, so these are uniquely determined. Junctions E and F then turn out to have labelling F1 and the labelling of junctions A and D is determined.

Holes are not connected to the outside lines, so the constraints are not propagated. Fig. 25 illustrates the search of a hole region. The label sets are found in the order of the letters from the T junction E. Line EF has a rightward

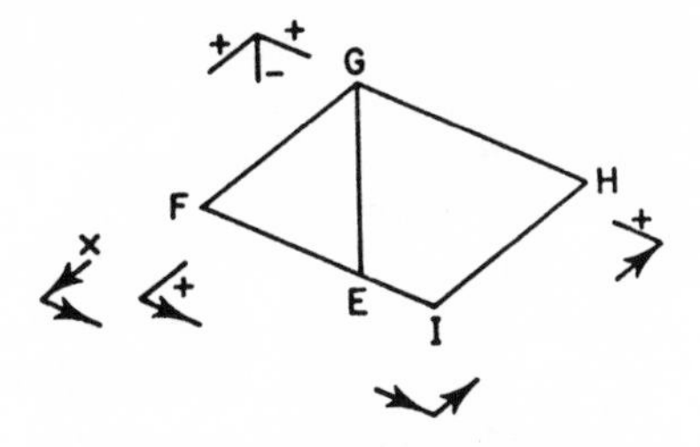

Fig. 25. Labelling of a hole region of a three-dimensional body. The cross indicates that a label has been removed from the set of possibilities

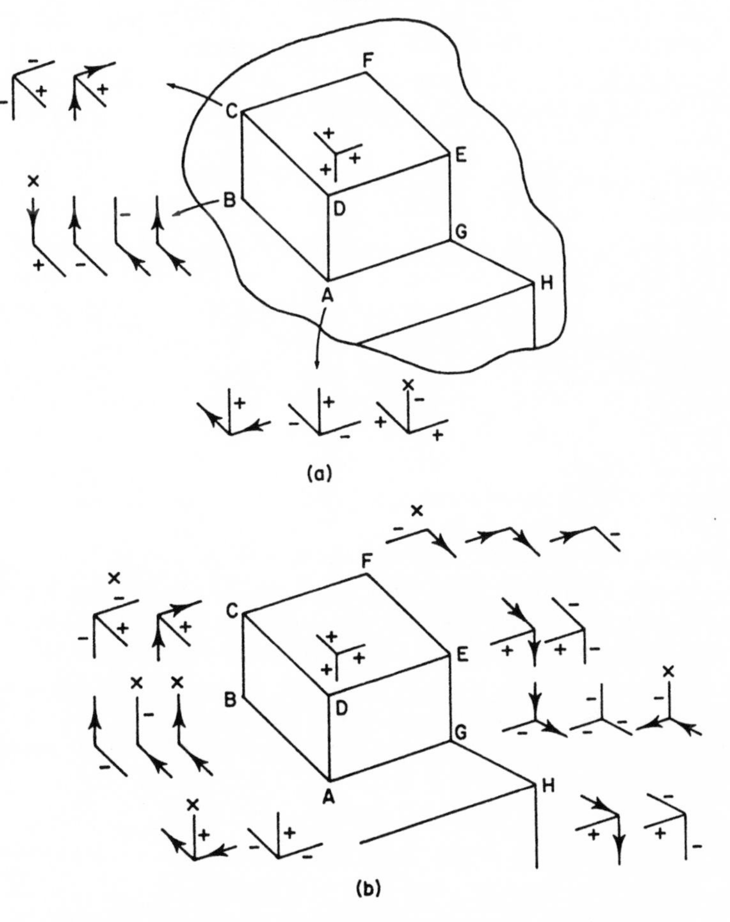

Fig. 26. Labelling a general line drawing, with (a) the search completed up to D, and (b) the search completed up to H

arrow. There are two possible labels for F, but when the search reaches G the labels are uniquely determined; so the labelling for F can only be L3. The labelling of H and I is then immediately determined.

If the drawing does not include outside lines (e.g. if they extend beyond the plane of the drawing), a more complicated search is necessary before the drawing can be correctly labelled. Fig. 26 shows an example of this. In this figure we cannot know in advance which boundaries are those of the polyhedron. We therefore start the search from the A junction, which has least labelling possibilities. Assume that we search this figure in the order of the letters, from A to H. When the search proceeds through A, B, and C, the label L4 of the junction B is eliminated by transmission of the constraint from C. Fig. 26(a) shows the situation when the search has got as far as D. The search is continued in the same way as far as H. No backwards propagation of the constraints occurs until the label set of H is determined. When this label set of H is determined, the result is propagated in order to G, A, B, C, and F so that the label sets for these junctions is updated.

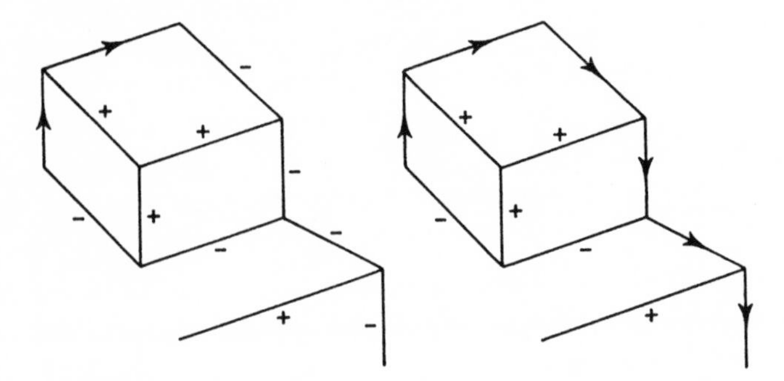

Fig. 27. Two possible interpretations of a line drawing

As a result, a single label is determined for A, B, C, and D. It can occur that two possible sets of labels are left for E, F, G, and H. As shown in Fig. 27, if E, F, G, and H are in contact with another face they would have the label '−'; otherwise they have the arrow label. Either of these would be a correct solution.

Thus a global interpretation can be arrived at by using local constraints. This technique can be applied to problems in various fields of artificial intelligence.

Chapter 4

Solving Problems by Decomposition

This chapter is concerned with the strategem of decomposing a given problem into several simpler sub-problems. If the problem can be broken down, the problem space can then be represented by an AND/OR graph; and methods of searching that can be applied to an AND/OR graph include searching by minimizing the cost and methods using heuristic knowledge.

The problem of deciding the next move in a game is a well-established part of the subject matter of artificial intelligence. Games can be represented by AND/OR graphs, so a strategem for deciding the next move in a game is to reduce it to the problem of searching such a graph.

4.1 DECOMPOSING PROBLEMS

In this section we shall describe the process of generating sentences as an example of how a problem can be simplified. The task is to generate sentences in conformity with a given grammar.

We shall assume that the grammar can be expressed by the following very simple rewriting rules:

R1 S → SUB PRED
R2 SUB → PRON
R3 SUB → NP
R4 PRED → V NP
R5 NP → DET N
R6 PRON → He
R7 V → saw
R8 DET → a
R9 N → dog
R10 N → cat

The items on the left (in capitals) are non-terminal symbols, and the items on the right of R6–R10 (in small letters) are terminal symbols. Starting from the initial state S, we wish to construct a sequence of symbols consisting only of terminal symbols, by repeated application of the rules. We shall express a sequence of symbols as a list in brackets.

By applying rule R1, which is applicable to the initial state S, we obtain (SUB PRED). The problem is now to convert (SUB PRED) into a sequence of terminal symbols. The sub-problems obtained by decomposing this problem are therefore to convert the list elements SUB and PRED into sequences of terminal symbols. These sub-problems are simpler than the original problem and can be solved independently. By applying R4 to PRED, for example, we obtain (V NP), and this can in turn be decomposed into V and NP. Fig. 28 shows an example of a procedure for obtaining one solution (there are others). To indicate that the problem has been decomposed, we put arcs between the edges of the tree.

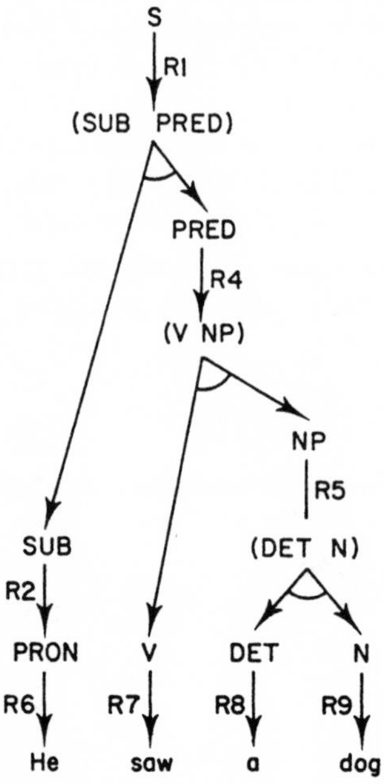

Fig. 28. Example of sentence generation problem

Now consider the cost of decomposing the problem. Assume that application of the rules takes time, and the cost is the time required to solve the whole problem. Let us represent the time taken to apply a single rule R by $t(\mathrm{R})$. If several rules can be applied independently, assume that they can be applied in parallel. If we let $f(n)$ be the cost of reaching a terminal symbol from a node n, we have $f(\mathrm{PRON}) = t(\mathrm{R6})$, $f(\mathrm{V}) = t(\mathrm{R7})$, $f(\mathrm{DET}) = t(\mathrm{R8})$, and $f(\mathrm{N}) = t(\mathrm{R9})$.

If we take it that n is resolved into child nodes $n_1, n_2, \ldots, n_m$, we have:

$$f(n) = \max\{f(n_1), f(n_2), \ldots, f(n_m)\} \tag{6}$$

Therefore $f(\mathrm{NP}) = t(\mathrm{R5}) + \max\{f(\mathrm{DET}) + f(\mathrm{N})\}$. By carrying out this calculation for all the nodes, we can find the total cost $f(\mathrm{S})$.

If the rules are applied in parallel, instead of equation (6) we must use the following:

$$f(n) = f(n_1) + f(n_2) + \ldots + f(n_m) \tag{7}$$

Equations (6) and (7) are typical examples of the cost calculations that are used when a problem is decomposed. One or other of these cost calculations is used in practically all problems.

Note that we have not considered the cost of decomposition. In Fig. 28, this cost can be included in the cost of the edge immediately before the node corresponding to the decomposition. For example, the cost of decomposition (SUB PRED) may be included in the cost of R1.

4.2 AND/OR GRAPHS

The sentence generation problem discussed in the previous section can be represented by the AND/OR graph of Fig. 29. In this case the graph is a tree.

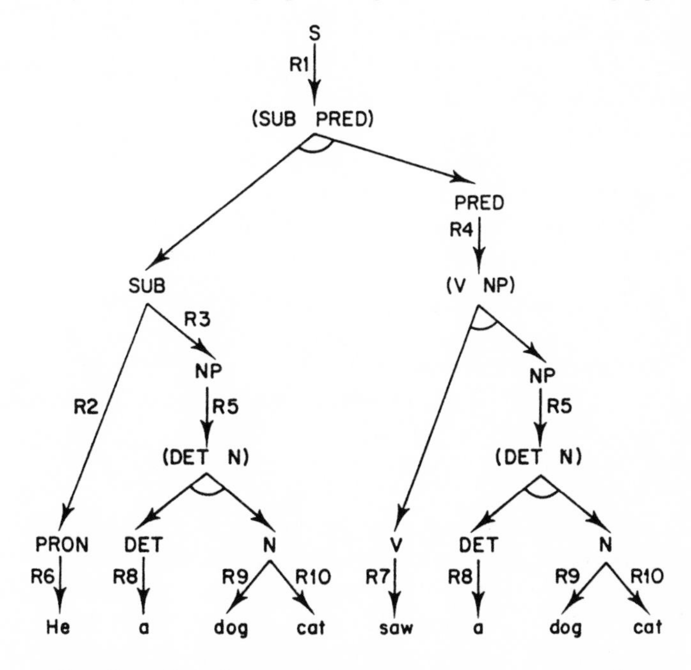

Fig. 29. AND/OR graph of sentence generation problem

In the state space graphs we dealt with in the earlier chapters it was sufficient to find a single goal node by following the edges. In an AND/OR graph there may be branched edges (representing the process of decomposing the problem into sub-problems). Nodes indicating such decomposition are called AND nodes. In Fig. 29 the nodes SUB and PRED are AND nodes. To solve the problem corresponding to the parent node of the AND nodes, we must solve the problems corresponding to all these AND nodes. In the figure the child nodes of the nodes that are not marked by arcs are called OR nodes. OR nodes have the same meaning as the nodes in the graphs discussed in the preceding chapters. That is, to solve the problem represented by the parent node of a number of OR nodes, it suffices to solve the problem represented by any one of such OR nodes.

Thus the solution of the problem has a structure such as is shown in Fig. 28; that is, the solution is part of an AND/OR graph. Such a partial graph is called a *solution graph*, and the drawing shows a problem in which all the end-points of the solution graph are solved. These end-points are called *terminal nodes*. Fig. 28 is the solution graph of the AND/OR graph of Fig. 29.

Solving a problem represented by an AND/OR graph involves searching all the solution graphs that it contains.

We may summarize the characteristics of AND/OR graphs and solution graphs as follows:

(1) Terminal nodes are solved nodes.
(2) If the child nodes of any given node are AND nodes, that node is solved when all these AND nodes are solved.
(3) If the child nodes of any given node are OR nodes, that node is solved when any one of these OR nodes are solved.
(4) A solution graph is a part of the AND/OR graph of a problem; which part is itself an AND/OR graph, all of whose nodes are solved, and in which no node, including the start node, has more than one edge to an OR node.

The AND/OR graph of Fig. 29 forms an AND/OR tree. If the two NP nodes in this figure are represented as a single node, it is still an AND/OR graph, but not an AND/OR tree. If we are only trying to find one solution of the problem, it is simpler to use an AND/OR graph. Here our intention in representing the node as two nodes was to obtain several types of solution. In fact, the solution 'a dog saw a cat' is also generated.

Fig. 30(a) shows that a complex problem A can be solved by decomposing it into B and C, or by solving D. Thus a graph in which both AND and OR nodes are present together at the same parent node can be converted into a graph in which AND and OR nodes are not mixed together in the child nodes, by providing an additional node as shown in Fig. 30(b). When cost is part of the problem, the cost of the edge from A to M must be regarded as the cost of

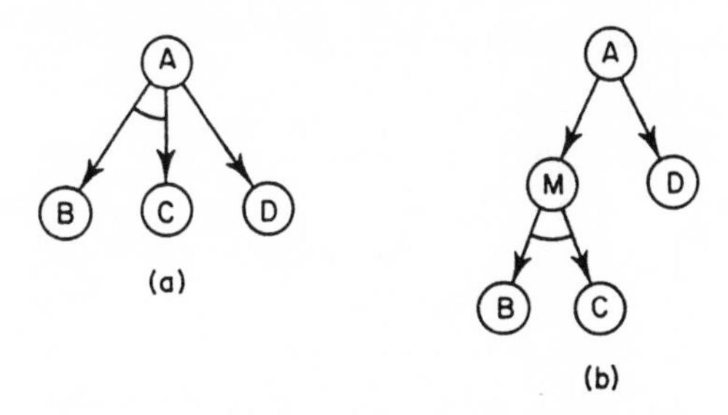

Fig. 30. Methods of representing mixed AND/OR nodes: (a) both AND and OR nodes as child nodes of same parent, and (b) AND/OR graph

breaking down A into B and C.

A node on an AND/OR graph may be an AND node of a given parent node, and, at the same time, an OR node of another parent node. It is therefore only when the parent–child relationship of the nodes is given that we are able to determine which it is. In this book we shall use the expressions 'n_i is an AND node of n' and 'the child node n_i of n is an AND node'.

4.3 SEARCHING AN AND/OR GRAPH

We shall now discuss various methods of finding solution graphs from an AND/OR graph. When searching a simple graph it was sufficient to find a single path; so, if we had pointers from the nodes that had been searched to the parent nodes, once the search was completed we could find the path constituting the solution. However, a solution graph of an AND/OR graph is itself a graph. The process of finding solution graphs by searching corresponds to progressively developing these solution graphs.

Only some of the possible solution graphs are found during the search. These are called *unsolved graphs* (they may also be called *partial-solution graphs*, but since we do not yet know whether they really will form part of a solution graph, we shall call them unsolved graphs).

Usually several unsolved graphs are found during the search. The structures of these partial graphs must therefore be represented in the computer.

4.3.1 Evaluation and expansion of unsolved graphs

An AND/OR graph can be thought of as repeated evaluations and expansions as follows:

1. Find whether the nodes of the unsolved graphs are solved, insoluble, or undecidable. If required, find the cost of the nodes.

2. Expand the unsolved graphs.

The processing which these involve is described below.

In step 1, first of all examine the end-points (nodes which have no child nodes) of the unsolved graphs. If they are terminal nodes, they correspond to a solved problem, so give them a mark indicating that they are solved. (We shall call this the SOLVED mark.) If the end-points of the unsolved graph are end-points of the original problem but are not terminal nodes, the problems corresponding to these nodes cannot be solved, so give them the mark UNSOLVED. If the child nodes of nodes other than end-points of unsolved graphs are AND nodes, and if all these child nodes are SOLVED, give such nodes also the SOLVED mark; but if even one child node is UNSOLVED, give the parent node the UNSOLVED mark. If the child nodes of a node other than the end-points are OR nodes, and have already been given either kind of mark, give their parent node the same mark.

If the start node S of an unsolved graph becomes SOLVED, the graph is a solution graph. If S becomes UNSOLVED, the graph cannot become a solution graph.

If it is necessary to calculate the cost, we find the cost of all the nodes as follows, using the costs of the end-points of the unsolved graphs. The cost of a parent node of AND nodes is computed, by means of equation (6) or (7), from the cost of all its child nodes. The cost of a parent node of OR nodes is the sum of the cost of the OR nodes and the cost of the edges. The cost of the start node is the cost of the unsolved graph.

In step 2, we find the child nodes by expanding the nodes which are end-points of the unsolved graph p and have not yet been given any mark. If the child nodes of an expanded node n are AND nodes, a new unsolved graph is generated by adding all the child nodes to p. If the child nodes of n are OR nodes, adding one only of these child nodes to p will generate as many unsolved graphs as there are such child nodes. In this way, a large number of unsolved graphs are generated in the searching process.

4.3.2 Depth-first search of an AND/OR graph

As in the case of searching an ordinary graph, the strategies available for searching an AND/OR graph are depth-first searching, breadth-first searching, and searching for an optimal solution. The search algorithm for an AND/OR graph may be expressed by the following procedure:

Procedure *depth-first AND/OR*

1. Put into the *open* list an unsolved graph consisting only of the start node.
2. *LOOP*: **if** *open* = empty **then** *exit* (*fail*).
3. *p*: = *first* (*open*); fetch the first element (unsolved graph) of the *open* list.

4. **If** *solved* (*p*) **then** *exit* (*success*); if *p* is a solved graph, the search has succeeded and *p* is a solution.
5. *Remove* (*p*, *open*).
6. Expand *p*, evaluate all the newly generated unsolved graphs, and put any unsolved graphs whose start nodes are not *UNSOLVED*, at the head of the *open* list.
7. **Goto** *LOOP*.

Table 5 shows the unsolved graphs in the open list that are obtained by applying this procedure to the AND/OR graph of Fig. 31. When the first unsolved graph on the *open* list is expanded and evaluated on the third repetition of this procedure, it is found to be UNSOLVED, and so has to be removed from the *open* list. On the sixth repetition of the procedure, the unsolved graph fetched from the *open* list is SOLVED, so it now becomes a solution graph and the search terminates.

TABLE 5
Depth-first search of Fig. 31

Number of times procedure repeated	Unsolved graphs in open list	
1	S	
2	S, A	S, B
3	S, A, C, D	S, B
4	S, B	
5	S, B, E	S, B, H
6	S, B, E, t_3, t_4	S, B, H

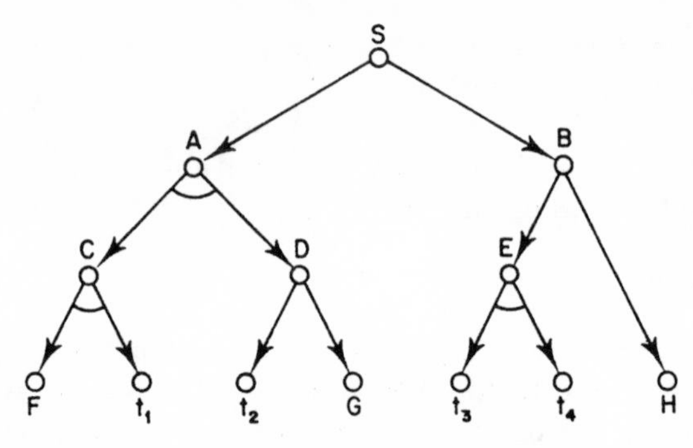

Fig. 31. Searching an AND/OR graph. t_1–t_4 are terminal nodes

In this procedure, the expanded unsolved graphs are stored in the *open* list, but it is not essential to do this in the case of a depth-first search. It is also possible to use a system in which this list only contains one partial graph, and if it is found that this cannot form a solution graph, the computer tries to find another solution graph.

To realize this system, it is necessary to provide a function such that, when the nodes are expanded, instead of all the child nodes being generated at once, only one child node is generated at a time. For this purpose we define the function *nextchild*. Let us assume that the node n has child nodes n_1, n_2, ..., n_m, and that the child node n_i has already been generated. (If no child node has yet been generated, $i = 0$). If we now apply the function *nextchild* to the node n, we obtain the following result:

$$\mathit{nextchild}(n) = \begin{cases} n_{i+1}, 0 \leqslant i < m \\ \mathrm{NIL}, i = m \end{cases}$$

A depth-first search of an AND/OR graph G using this function can be represented by the following procedure:

Procedure *depth-first AND/OR-1*

1. Construct unsolved graph p consisting only of the start node. n := S. Evaluate p.
2. *LOOP*: **if** *unsolved* (p) **then** *exit* (*fail*).
   If *solved* (p) **then** *exit* (*success*).
3. **If** n is marked, **goto** *EVALUATE*.
4. *nextn* := *nextchild* (n).
   Add *nextn* to p, evaluate p.
   n := *nextn*.
   Goto *LOOP*.
5. *EVALUATE*: evaluate p.
   If n is *UNSOLVED*, remove n from p.
   n := parent node of n.

Goto *LOOP*.

The end-point n of the unsolved graph changes as follows when this procedure is applied to the AND/OR graph of Fig. 31: S → A → C → F → C → A → S → B → E → t_3 → E → t_4. This searching sequence is the same as the depth-first search of an ordinary graph.

If, during the progress of the search, we find that any OR node is UNSOLVED, we return to the immediately preceding node and search another edge. This process is called *backtrack*. Because the depth-first AND/OR-1 procedure performs this backtrack function, there is no need (as there is in the depth-first AND/OR procedure), to expand all the child nodes of a node and store them in the open list.

In the depth-first AND/OR procedure, when an unsolved graph p was expanded, if the child nodes of the end-point n of p are AND nodes, we add all the child nodes to p. If we want to reduce the number of times we find child nodes, we can find them one-at-a-time, and evaluate each one as it is found. If an UNSOLVED child node is found, n also becomes UNSOLVED and there is no need to find further child nodes. To carry out such a process, it is useful to use the function *nextchild*.

4.3.3 Searching for optimal solution of AND/OR graph

We now deal with the problem of finding the solution graph for which the total cost is a minimum when the costs on edges going to OR nodes of the AND/OR graph are defined. Let us assume that for each node of the graph, the estimated cost which its solution will require is given. These correspond to $\hat{h}$ in the method of searching an ordinary graph explained in Section 3.4. So let us call the estimated cost of searching a node n, $\hat{h}(n)$. If no heuristic function $\hat{h}(n)$ is given, this is the same as $\hat{h}(n) \equiv 0$. Putting $\hat{h}(n) \equiv 0$ in the following algorithm turns it into a breadth-first search of the AND/OR graph.

As in the depth-first AND/OR described in the preceding section, in the search for an optimum solution of an AND/OR graph various unsolved graphs are generated. The inferred value $\hat{f}(p)$ of the cost of the unsolved graph p is the inferred value of the cost obtained by evaluating p using the inferred value of the cost of the end-points. Fig. 32 shows part of an AND/OR graph,

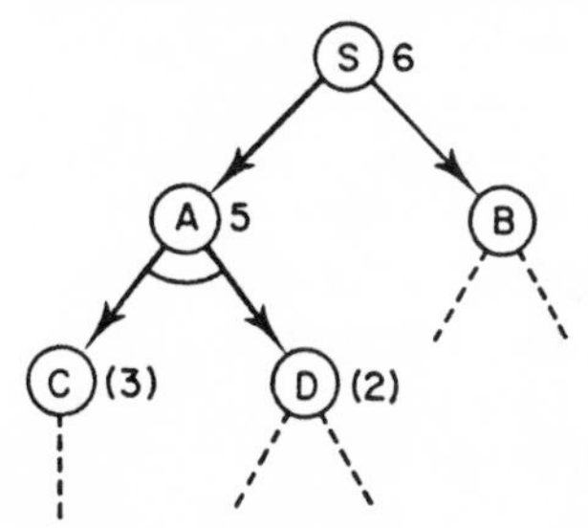

Fig. 32. Estimating the cost of the unsolved graph. Figures in brackets are $\hat{h}$, and figures without brackets are evaluated costs

and an unsolved graph p shown by the edges in heavy lines. The costs of the AND nodes are found using equation (7), and the costs of the edges leading to OR nodes are all taken to be 1. If the estimated values of the costs of the nodes A and S are calculated using the inferred values $\hat{h}(C) = 3$ and $\hat{h}(D) = 2$ of the costs of the end-points C and D of p, we have $\hat{f}(A) = \hat{h}(C) + \hat{h}(D)$ and $\hat{f}(S) = \hat{f}(A) + 1$. Therefore $\hat{f}(p) = 6$.

As with the A-algorithm, the search proceeds by expanding the unsolved graph of minimum $\hat{f}(p)$. This algorithm can be expressed by the following procedure:

Procedure *optimal AND/OR*

1. Put an unsolved graph p consisting only of the start node S in the open list. $\hat{f}(p) := \hat{h}(S)$.
2. *LOOP*: **if** *open* = empty **then** *exit*(*fail*).
3. p :=*first*(*open*).
4. **If** *solved* (p) **then** *exit*(*success*).
5. *Remove* (p,*open*).
6. Expand p. Evaluate all unsolved graphs p_i. that are newly generated. Find $\hat{f}(p_i)$. If p_i is UNSOLVED, put it in the *open* list. List the unsolved graphs of the *open* list in increasing order of $\hat{f}$.
7. **Goto** *LOOP*.

Table 6 shows some of the changes of the unsolved graphs of the *open* list when the AND/OR graph of Fig. 33 is searched by the optimal AND/OR algorithm. The figures in brackets next to the nodes are the estimated costs, and the figures without brackets are the costs calculated using these. The unsolved graphs are named, to avoid having to draw the same graph more than once.

The table shows only the first six cycles. In the subsequent cycles, H and I of $p8$ are expanded and thus terminal nodes t_1, and t_2 are added on, to obtain the solution graph $p10$. (The cost $f(p10)$ of $p10$ is 5.)

Even in this procedure, if the child nodes of a node n of an unsolved graph p which we are trying to expand are AND nodes, they may be evaluated by generating them one-at-a-time instead of all at once. Calling the child nodes n_1, $n_2, \ldots, n_m$, if the procedure has expanded up to n, when p is expanded we find only n_{i+1}. If n is UNSOLVED, the unsolved graph which is generated is also UNSOLVED, and so need not be put on the *open* list. Also, even in cases other than this, if we take $\hat{f}(n)$ as $\hat{f}(n_1) + \hat{f}(n_2) + \ldots + \hat{f}(n_{i+1})$, if $\hat{f}(n)$ has a large value it may not be fetched again after being put in the *open* list. If it is not fetched, we need only generate a child node a few times.

As was the case with a graph search carried out using the ordinary A-algorithm, there is no guarantee that this algorithm will find the optimal solution. However, we can prove (though the proof is omitted here) that the optimal solution can always be found if $\hat{h} \leqslant h$ for all nodes.

TABLE 6
Optimal-solution search of Fig. 33

Number of cycles	Unsolved graphs in open list		
1	p1  S (4)		
2	p2 S 4, A (3) 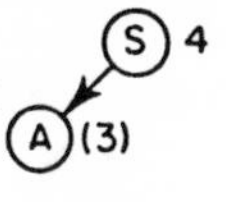	p3 S 4, B 3 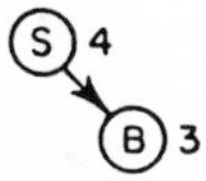	
3	p3	p4 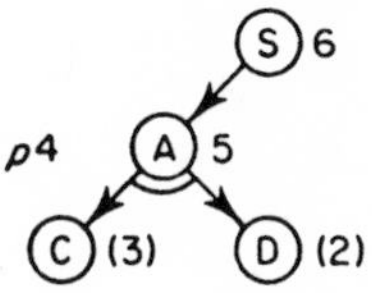 S 6, A 5, C (3), D (2)	
4	p5 S 4, B 3, D (2), I (1) 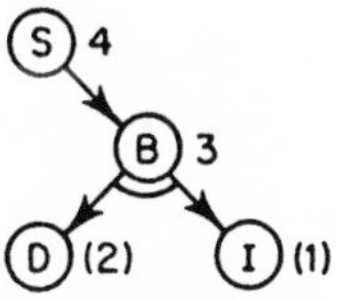	p4	
5	p6 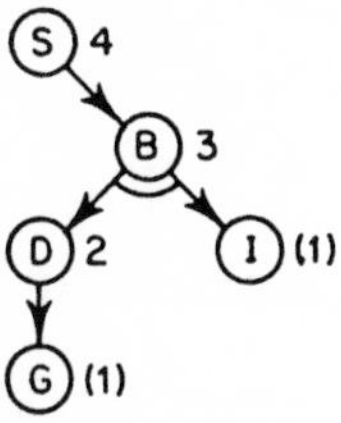 S 4, B 3, D 2, I (1), G (1)	p7 S 5, B 4, D 3, I (1), F (2) 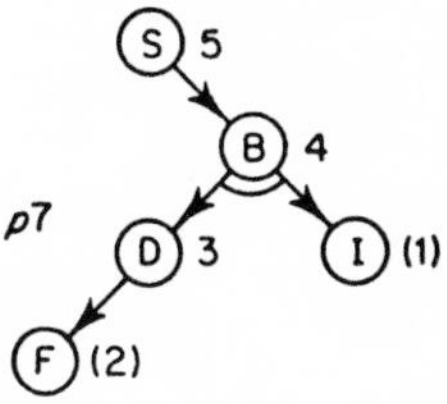	p4
6	p8 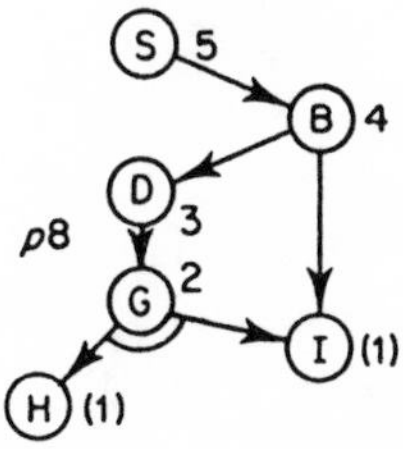 S 5, B 4, D 3, G 2, I (1), H (1)	p7	p4

If, in calculating the inferred costs of the unsolved graphs, we set the cost of all edges equal to 0, this corresponds to trying to minimize future searching costs.

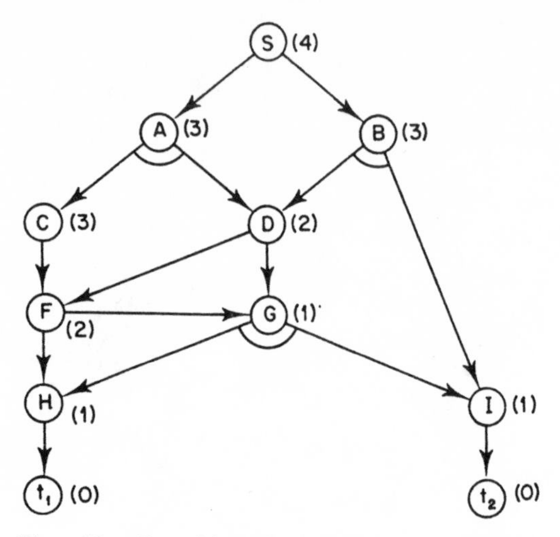

Fig. 33. Searching for optimal solution of AND/OR graph

4.4 SEARCHING A GAME TREE

4.4.1 Game tree

The problem of deciding the next move in the game can be solved using the search techniques we have described. We assume that the two players move alternately (i.e. the operators are applied alternately), starting from the initial state of the game until the final game state is reached, whereupon the final state is evaluated by giving it an evaluation value. If the final state can only be a win, a loss, or a draw, the evaluation value may be a suitable score corresponding to these outcomes. Various states are produced during the progress of the game. These can be represented by means of a graph. Sometimes the same state may be reached by different sequences of moves; but if we ignore this and represent it by different nodes, the graph will still be a tree. In this chapter we shall represent games by trees, for simplicity.

Assume that the goal of the player to move first (called MAX) is to obtain the maximum evaluation value and the goal of the player to move next (called MIN) is to obtain the lowest evaluation value, and that both players choose the best moves. Fig. 34 shows an example of a simple game tree. Nodes

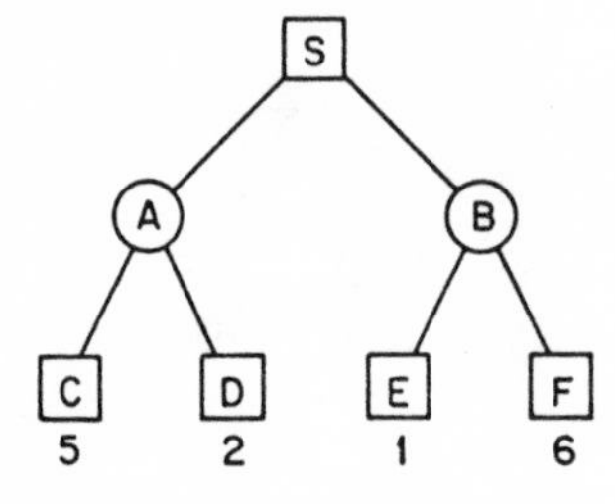

Fig. 34. Simple game tree

corresponding to MAX's moves are shown as squares, and nodes corresponding to MIN's moves are shown as circles. The evaluation values are given at the end-points.

In normal games the tree is of greater depth and there are more branches from each node. For games of even moderate complexity, it is in practice impossible to generate a tree including all the end-points. In such cases, the evaluation values are found by generating a tree of suitable depth, and evaluating the end-points of this partial tree by some method. The values determined in this way are called *static evaluation values*. If the game is near its end, we can reach the true end-points of the game tree; but otherwise in most cases the best move must be decided using the static evaluation values. We take it that when the opponent determines his move he will again play the best move, after generating a partial tree of suitable depth. Such a look-ahead process is effective because the lower nodes are closer to the final state than the upper nodes, and so give a more accurate evaluation of the state of play.

The depth of the look-ahead is restricted by the amount of computing time which this requires. The static evaluation values must be determined using knowledge about the characteristics of the game. To play games such as chess, Shogi, or Go, we need to investigate not only methods of searching the game tree, but also suitable methods of evaluation.

However, in this chapter we shall not discuss the problem of evaluation, but will simply describe methods of searching when the evaluation values of the nodes are known.

4.4.2 Minimax method

We shall explain the *minimax* method with reference to the example of Fig. 34. From the start node S, MAX has two possible nodes. If we take the first edge, we come to node A. MIN then has two moves to choose from. MIN, of course, chooses the move (right-hand edge) that leads to the node D of smallest evaluation value. If MAX had chosen node B initially, MIN would choose node E, giving an evaluation value of 1. MAX must choose his own moves assuming that the evaluation value at each MIN node (circled node) is the lowest of the evaluation values of all the child nodes leading from it. If we call the MAX node n and its child node n_i (i = 1–m), and the child node of each n_{ij} ($j = 1 - i_m$), MAX maximizes the evaluation value of $f(n_i)$. That is, he selects i so as to satisfy the following:

$$f(n_i) = \max_i f(n_i)$$

MIN minimizes $f(n_i)$ for the respective i, and so chooses j so as to satisfy the following:

$$f(n_{ij}) = \min_j f(n_{ij})$$

So MAX chooses i in accordance with the following:

$$f(n_{ij}) = \max_i \{\min_j f(n_{ij})\}$$

If we take the depth of the tree as k, MAX should select i_1 such that:

$$f(n_{i_1 i_2 \ldots i_k}) = \max_{i_1} \min_{i_2} \ldots \{f(n_{i_1 i_2 \ldots i_k})\}\,.$$

This is called the minimax strategy.

If the evaluation values of the end-points of the game tree are correct, the minimax method will select the best move. However, if the depth of a tree increases, the number of nodes that must be examined increases exponentially. Research has therefore been directed towards finding more efficient algorithms. A typical example is explained below.

4.4.3 Alpha–beta method

Consider searching the game tree shown in Fig. 35. The nodes are numbered and a score is given for each of the terminal nodes. If we find the best moves for both sides using the minimax algorithm, initially MAX chooses node 1, then MIN chooses node 3, then MAX chooses node 8, which has an evaluation value of 5.

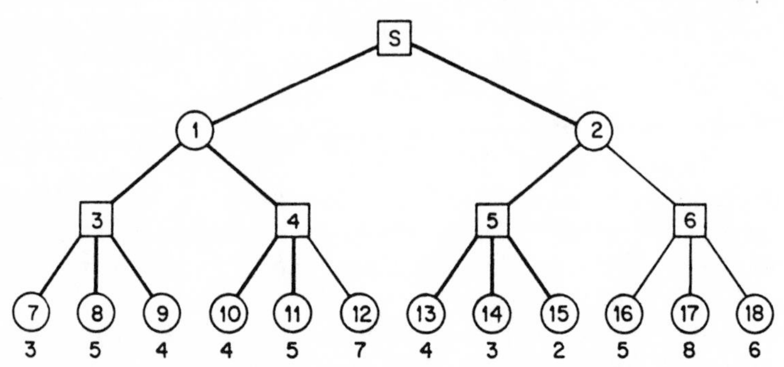

Fig. 35. Search of game tree using alpha–beta algorithm

Assume that, in a procedure which is the same as a depth-first search, the evaluation value 5 has been obtained for node 3, and we are trying to find the value of node 4. At this time-point, we know that the value of node 1 cannot be greater than 5. (This is because, if the value of node 4 were greater than 5, MIN would not choose node 4.) As soon as we know that the values of nodes 10 and 11 are obtained, the value of node 4 is at least 5. So we now know that it cannot be more advantageous for MIN to choose node 4 rather than node 3. We may therefore conclude that MIN will choose node 3, even without examining node 12. In this way, the value of node 1 is determined to be 5, so we know that the value of S must be at least 5. Now consider the time-point at which an evaluation value of 4 is obtained for the node 5 by examining the values of all the child nodes of node 5. We now know that selecting node 2 will bring MAX no advantage. So there is no need to look at node 6 or nodes beyond it. Edges leading to nodes that are searched are shown by heavy lines in the figure. This method of cutting down the number of nodes that need to be searched is called the alpha–beta method.

For systematic searching of deep trees, this method is easier to understand if it is formalized as follows. Let the lower bound and upper bound of each node be respectively α and β. In the initial state, the α and β of S are respectively $-\infty$ and ∞. We write this as $\alpha\beta(S) = (-\infty, \infty)$. If a value x is below or equal to α or above or equal to β, we say that x is outside the range (α, β). When the value of the node 3 is determined, $\alpha\beta(1) = (-\infty, 5)$. If we know that the values of the child nodes of node 1 are outside the range $(-\infty, 5)$ (in this example, not more than 5), we can interrupt the search. When node 11 has been searched, $\alpha\beta(4) = (5, \infty)$. Since $(5, \infty)$ is outside the range $(-\infty, 5)$, the condition for interrupting the search is satisfied, so we need not search further child nodes of node 4. We now have $\alpha\beta(1) = (5, 5)$ and $\alpha\beta(S) = (5, \infty)$. As the search proceeds, at the time when the value of node 5 is found to be 4, $\alpha\beta(2) = (-\infty, 4)$, which is outside the range $(5, \infty)$, so we can interrupt the search of node 2 and nodes beyond it.

The first interruption is because the value of node 4 is equal to or greater than the value of β of its parent node; and the second interruption is because the value of node 2 is equal to or less than the value of α. These are called a β-*cut* and an α-*cut* respectively. This method of increasing the efficiency of the search using the lower and upper bounds of the nodes is known as the alpha–beta method.

To evaluate the performance of the method we shall find out how many nodes have to be searched in the best case. Consider the time when we have found the value θ of the first child node of MAX's node n. $\alpha\beta(n) = (\theta, \infty)$. If the β of the parent node m of n is less than or equal to θ, we need not examine further child nodes of n. This is always a possibility if β is not ∞. If the node m is not the first child node of the parent node of m, its β is already given, and so cannot be ∞. Similar conclusions can be drawn for the MIN nodes. Consequently, in the best case, if the node m is not the first node of its parent node, we do not need to examine any child nodes other than the first child node of m.

The terminal nodes are represented by the sequence of edges that lead to them. For instance node 12 of Fig. 35 may be expressed as (1,2,3), since it is reached by following the first edge from S, the second edge from node 1, and the third edge from node 4. Thus the terminal nodes of a tree of depth k are represented by $(e_1, e_2, \ldots, e_k)$. For nodes that need not be examined, we have:

$$e_i \neq 1, e_{i+1} \neq 1$$

Consequently, for nodes that have to be examined in the best case, either all the odd-numbered nodes in the sequence are 1 (these are called the *odd nodes*), or all the even-numbered nodes in the sequence are 1 (these are called the *even nodes*).

If we assume that the number of edges issuing from non-terminal nodes is always m, then the number of odd nodes $N_o(k, m)$ and the number of even nodes $N_e(k, m)$ are given by the following formulae:

when k is odd: $N_o(k, m) = m^{(k-1)/2}$, $N_e(k, m) = m^{(k+1)/2}$
when k is even: $N_o(k, m) = m^{k/2}$, $N_e(k, m) = m^{k/2}$

The node (1, 1, ..., 1) is counted by both $N_o(k, m)$ and $N_e(k, m)$, so the total $N_{best}(k,m)$ is $N_o(k, m) + N_e(k, m) - 1$. That is:

when k is odd: $N_{best}(k, m) = m^{(k-1)/2} + m^{(k+1)/2} - 1$
when k is even: $N_{best}(k, m) = 2m^{k/2} - 1$

In the best case, the number of nodes that must be examined with the alpha–beta method is approximately the same as the number of nodes that would have to be searched by the minimax method with a tree of half the depth. However, in the worst case all the nodes must be examined. Usually the situation is somewhere between these extremes; that is, it may be approximated by the following inequality:

$$m^{k/2} < N(k, m) \leqslant m^k$$

4.4.4 Using the method of searching an AND/OR tree

A game tree may be expressed by an AND/OR tree. For example, the game tree of Fig. 34 may be expressed by the AND/OR tree shown in Fig. 36. The child nodes of MAX nodes correspond to OR nodes, and the child nodes of MIN nodes correspond to AND nodes (the start node is taken as being an AND node). Unlike the AND/OR trees we have seen in the previous section, the terminal nodes are given evaluation values. The costs of the edges are always 0. In the minimax method, the best move is determined by, if the child nodes of a given node are AND nodes, taking the least (evaluation) value of the child nodes as the value of that node; and, if the child nodes are OR nodes, taking the maximum value of the values of the child nodes as the value of that node.

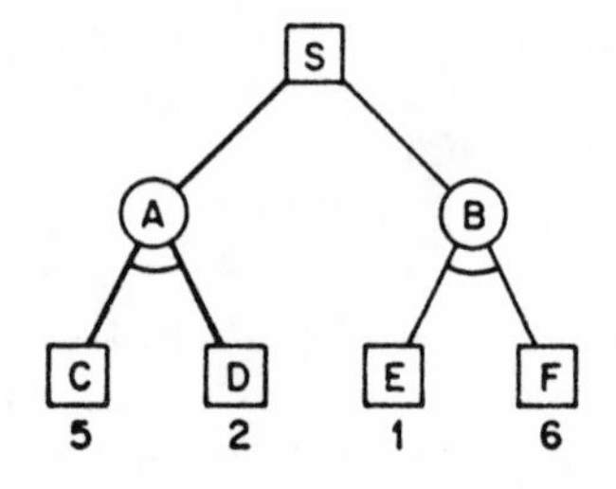

Fig. 36. AND/OR game tree

In the case of a game, the goal is to make the value of the start node a maximum. (The situation regarding evaluation values is the opposite to that regarding cost—the larger the evaluation value the better.)

Consequently, the method of computing the values of the various nodes of the solution graph (in this case the solution tree) is different from that used in the method of evaluation of an AND/OR tree described in the previous section. The evaluation value of the terminal node of the solution tree is itself the value of the terminal node. The value of a node whose child nodes are AND nodes is the minimum value of the values of the child nodes. The value of a node whose child node is an OR node is the same as the value of the child node. (Remember that there is only one child node in such a case.) The edges of one solution tree are shown by thick lines in Fig. 36. If we write the value of the node n as $f(n)$, $f(\mathrm{A}) = 2$ and $f(\mathrm{S}) = 2$. There is another solution tree apart from the one shown by thick lines. However, its value is 1, so the former is the optimal solution.

The AND/OR graph optimal-solution search method described in the previous section can also be used for searching an AND/OR game tree. However, $\hat{h}(=\hat{f})$ is not given beforehand except for the terminal nodes. To obtain the optimum solution we have to have $\hat{f}(n) \geqslant f(n)$ for all nodes. (When we were trying to obtain the minimum-cost solution, we had to have $\hat{h}(n) \leqslant h(n)$ for all nodes.) Consequently if the values of the end-points are not given, the inferred values of the unsolved trees are infinitely large. Let us assume that, when expanding non-terminal nodes of an unsolved tree, if the child nodes are AND nodes we expand them one at a time. If we use n_1 to represent the first child node (AND node) of n_i, we can see that, at the time when the value of n_i is found, $f(n) \leqslant f(n_1)$. So we may take $\hat{f}(n) = f(n_1)$. Thus we can calculate $\hat{f}$ and update it during the progress of the search. Like the optimal AND/OR search described in the previous section, the principle of this search is that the unsolved tree which is fetched from the *open* list to be expanded is always the one which has the largest inferred value.

When obtaining the optimum solution of an AND/OR game tree, instead of finding the solution in the form of a solution tree we could find only the terminal node that gives the value of the solution tree. This is because, since we are not dealing with a graph, once we know the terminal node we have only to follow it back to reach the start node. The algorithm is given below as a procedure. As in the previous section, once the value of the node n that is being searched is determined, we give n the SOLVED mark. For convenience, we give nodes that are not SOLVED the LIVE mark, and express the state of n by (n, s, v). s indicates whether the node is SOLVED or LIVE, and v gives $f(n)$ if s is SOLVED or $\hat{f}(n)$ if it is LIVE. The state of the node n is put onto the *open* list by the procedure *putopen*(n, s, v). We shall also write finding the parent node of n as the function *parent*(n); finding whether n is a terminal node or not as *terminal*(n); and finding whether n is an AND node or an OR node as *type*(n).

Procedure *game-tree search*

1. *Putopen* $(S, LIVE, \infty)$.
2. *LOOP*: **if** *open* = empty **then** *exit(fail)*.
3. $n := first(open)$.
4. **If** $(n = S \wedge solved\ (n))$ **then** *exit* (*success*).
5. *Remove*$(n, open)$.
6. Execute the processing shown in Table 7 in accordance with the state of n and the position of n in the AND/OR tree. List the nodes of *open* in order of largest $\hat{f}$. If the $\hat{f}$ values are the same, put the *SOLVED* node first.
7. **Goto** *LOOP*.

TABLE 7
Procedures of step 6 of the game-tree search

Procedure number	*Condition of n*	*Procedure*
1	$s=LIVE$ $type(n)=AND$ not $terminal(n)$	For all child nodes n_i of n, $putopen(n_i, LIVE, \hat{f}(n))$
2	$s=LIVE$ $type(n)=OR$ not $terminal(n)$	$nextn:=nextchild(n)$ **if** $nextn=NIL$ **then** $putopen(n, SOLVED), \hat{f}(n))$ **else** $putopen(nextn, LIVE, \hat{f}(n))$
3	$s=LIVE$ $terminal(n)$	$putopen(n, SOLVED, min\{\hat{f}(n), f(n)\})$
4	$s=SOLVED$ $type(n)=AND$	$putopen(parent(n), LIVE, \hat{f}(n))$
5	$s=SOLVED$ $type(n)=OR$	$m=parent(n)$ Remove all child nodes of m in *open* $putopen(m, SOLVED, \hat{f}(n))$

The changes in the *open* list produced by applying the procedure game-tree search to the AND/OR game tree of Fig. 37 are shown in Table 8.

Fig. 37 shows the optimum solution obtained by the search. The heavy line shows the nodes which have been searched. (The nodes marked LIVE left in the *open* list have still not been searched.) To find the path of the optimum solution, pointers to the parent nodes are set in procedures 1 and 2 shown in Table 7 when the child nodes are found. Since nodes which are removed from the *open* list in procedure 5 of Table 7 and any nodes n that satisfy $\hat{f}(n) \leqslant f(n)$ in procedure 3 of this table are not on the path of this optimum solution, their pointers may be removed. Thus, when the solution is obtained by this algorithm the path of the solution is obtained by the pointers.

For comparison, the terminal nodes which are searched when the alpha–beta method is applied to the same AND/OR tree are indicated by crosses in Fig. 37. It can be seen that game-tree search finds the solution in far fewer searching steps.

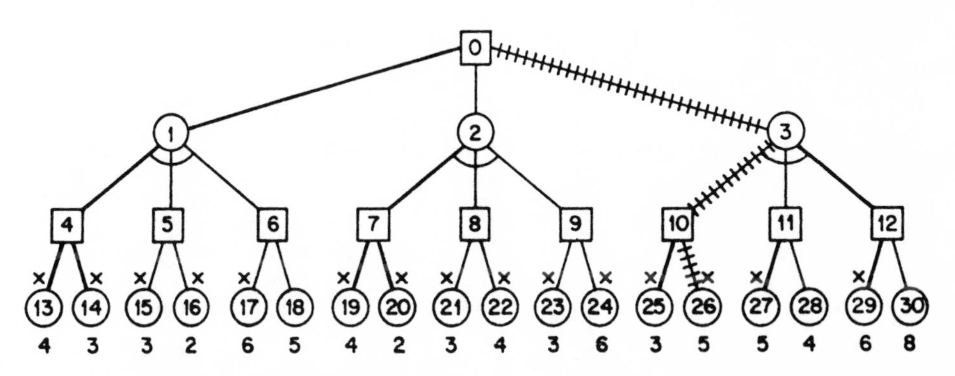

Fig. 37. Search for optimal solution of AND/OR game tree

However, if the partial AND/OR tree represented by node 1 and the nodes below it is interchanged in position with the partial AND/OR tree represented by node 3 and the nodes below it, the alpha–beta method would not need to search any more nodes than the game-tree search. Since the alpha–beta method uses a depth-first search to examine the nodes, its searching efficiency depends on the position of the nodes.

The optimal-solution search looks first at the most promising nodes, so its efficiency is not affected even if the arrangement of the nodes is changed to some degree. It has been proved that the nodes which are examined by the optimal solution search will always be examined by the alpha–beta method too.‡ That is, this method is more efficient that the alpha–beta method.

However, with this method several candidate solutions must be stored in the *open* list. With the alpha–beta method, since we need only know the values of α

‡Stockman, G. C., 'A minimax algorithm better than alpha–beta?', *Artificial Intelligence*, **12**, 1979, 179–196.

TABLE 8
Sequence for obtaining the optimum solution of the AND/OR game tree of Fig. 37

Turn number	*Number of procedure applied*	*Open list* (start node represented 0, *SOLVED* by S, and *LIVE* by L)
1		(0 L ∞)
2	1	(1 L ∞), (2 L ∞), (3 L ∞)
3	2	(4 L ∞), (2 L ∞), (3 L ∞)
4	1	(13 L ∞), (14 L ∞), (2 L ∞), (3 L ∞)
5	3	(14 L ∞), (2 L ∞), (3 L ∞), (13 S 4)
6	3	(2 L ∞), (3 L ∞), (13 S 4), (14 S 3)
7	2	(7 L ∞), (3 L ∞), (13 S 4), (14 S 3)
8	1	(19 L ∞), (20 L ∞), (3 L ∞), (13 S 4), (14 S 3)
9	3	(20 L ∞), (3 L ∞), (13 S 4), (19 S 4), (14 S 3)
10	3	(3 L ∞), (13 S 4), (19 S 4), (14 S 3), (20 S 2)
11	2	(10 L ∞), (13 S 4), (19 S 4), (14 S 3), (20 S 2)
12	1	(25 L ∞), (26 L ∞), (13 S 4), (19 S 4), (14 S 3), (20 S 2)
13	3	(26 L ∞), (13 S 4), (19 S 4), (14 S 3), (25 S 3), (20 S 2)
14	3	(26 S 5), (13 S 4), (19 S 4), (14 S 3), (25 S 3), (20 S 2)
15	5	(10 S 5), (13 S 4), (19 S 4), (14 S 3), (20 S 2)
16	4	(3 L 5), (13 S 4), (19 S 4), (14 S 3), (20 S 2)
17	2	(11 L 5), (13 S 4), (19 S 4), (14 S 3), (20 S 2)
18	1	(27 L 5), (28 L 5), (13 S 4), (19 S 4), (14 S 3), (20 S 2)
19	3	(27 S 5), (28 L 5), (13 S 4), (19 S 4), (14 S 3), (20 S 2)
20	5	(11 S 5), (13 S 4), (19 S 4), (14 S 3), (20 S 2)
21	4	(3 L 5), (13 S 4), (19 S 4), (14 S 3), (20 S 2)
22	2	(12 L 5), (13 S 4), (19 S 4), (14 S 3), (20 S 2)
23	1	(29 L 5), (30 L 5), (13 S 4), (19 S 4), (14 S 3), (20 S 2)
24	3	(29 S 5), (30 L 5), (13 S 4), (19 S 4), (14 S 3), (20 S 2)
25	5	(12 S 5), (13 S 4), (19 S 4), (14 S 3), (20 S 2)
26	4	(3 L 5), (13 S 4), (19 S 4), (14 S 3), (20 S 2)
27	2	(3 S 5), (13 S 4), (19 S 4), (4 S 3), (20 S 2)
28	5	(S S 5), (13 S 4), (19 S 4), (4 S 3), (20 S 2)

and β, less memory is needed for the computation. A suitable choice between these two methods must therefore be made depending on whether processing speed or memory capacity is most important.

Chapter 5
Control of Problem-solving

We have explained the basic methods of searching the state space of a given problem. The problems in question have been represented by comparatively simple state spaces. The knowledge used to perform an efficient search has been given in the form of the cost of specified states and the cost to the goal state, or the cost from the initial state to a specified state.

When a problem becomes complex it is difficult to reduce it to such simple search problems. How the strategy in solving the problem is to be determined becomes an important question. In this chapter we describe techniques for controlling the solution of such complex problems. First, to show the importance of controlling the solution of a problem, we shall describe a typical system, called 'generalized problem solver' (GPS). We shall then describe methods of problem-solving using predicate calculus, which is a logical formalization of the problem-solving process. We shall also explain methods of solving complicated problems by forming a rough plan.

5.1 IMPORTANCE OF CONTROL

People have to solve various problems in daily life. Most of them are solved sub-consciously, from experience. Take, for example, the problem of getting to a particular place by walking. To get there many actions are necessary, such as climbing stairs, opening doors, or choosing a path. A person has a great number of operators. Looking at a map, learning the condition of the surroundings, or making a telephone call are all such operators. Also, he has a lot of experience. Even if there were a robot with the same operators as a person and sufficient memory capacity to store the experience, it is doubtful whether it could solve problems as a person does. A given problem seldom coincides exactly with previous experience. There must therefore be the ability to select *similar* experience. It is also necessary to know how this experience should be altered to solve the present problem. This requires a knowledge of the effects of applying many operators, and how they can be suitably combined.

With the maze problem, or simple games, we had a uniform state space, so it was sufficient to search always towards the goal. In solving a complicated problem it is more efficient to set up a general plan and to attain, in sequence, several subgoals in the plan. This question of deciding what should be done next is an important problem in proving theorems, understanding natural languages, and image comprehension, etc., and is called 'control strategy of problem solving process' or 'control of focus of attention'.

5.2 *GENERAL PROBLEM-SOLVER*

General problem solver (GPS) is a model of the human problem-solving process which was proposed by the psychologists A. Newell, J. C. Shaw and H. A. Simon. This model provides a framework, not just for solving particular problems, but for solving arbitrary problems by using given knowledge about the problems. We shall see how GPS solves problems by taking a simple specific example.

5.2.1 Representation of problems by GPS

As an example we shall show how the 'monkey and banana' problem is represented in GPS.

Initial state

The monkey is at place a. The box is at place b. The monkey is not holding anything. We express this state of the world as follows:

AT(monkey, a)
AT(box, b)
EMPTY

Goal state

The monkey is holding the banana. This is expressed as follows:

HOLD(banana)

Operators

GOTO (u); the monkey goes to place u.
Preconditions: none
Delete list: AT(monkey, x); x can be anything
Add list: AT(monkey, u)
MOVEBOX (v); the monkey carries the box to place v.

Precondition: AT(monkey, x) ∧ AT(box, x); the monkey and the box are at the same place.
Delete list: AT(monkey, x), AT(box, x)
Add list: AT(monkey, v), AT(box, v)

CLIMB; the monkey climbs on the box.
Precondition: AT(monkey, x) ∧ AT(box, x)
Delete list: AT(monkey, x)
Add list: ON(monkey, box)

GRASP (banana); grasping the banana.
Precondition: AT(box, c) ∧ ON(monkey, box)
Delete list: EMPTY
Add list: HOLD(banana)

Differences

Three differences between states are involved in this problem:

D_1 difference of the position of the monkey
D_2 difference of the position of the box
D_3 difference of the state of the monkey's paw

The order of importance of differences is assumed given beforehand. In this example, the order of importance is D_3, D_2, D_1.

Relationship between differences and operators

Candidate operators to reduce the differences are given in a table (means-ends table). The relationship between the operators and differences shown above is set out in Table 9. This table is used to find which operators are effective to reduce a specified difference and move from a given state to another state.

TABLE 9
Relationship between the operators and the differences affected by them

Operators	*Differences* D_1	D_2	D_3
GOTO	•		
MOVEBOX	•	•	
CLIMB	•		
GRASP			•

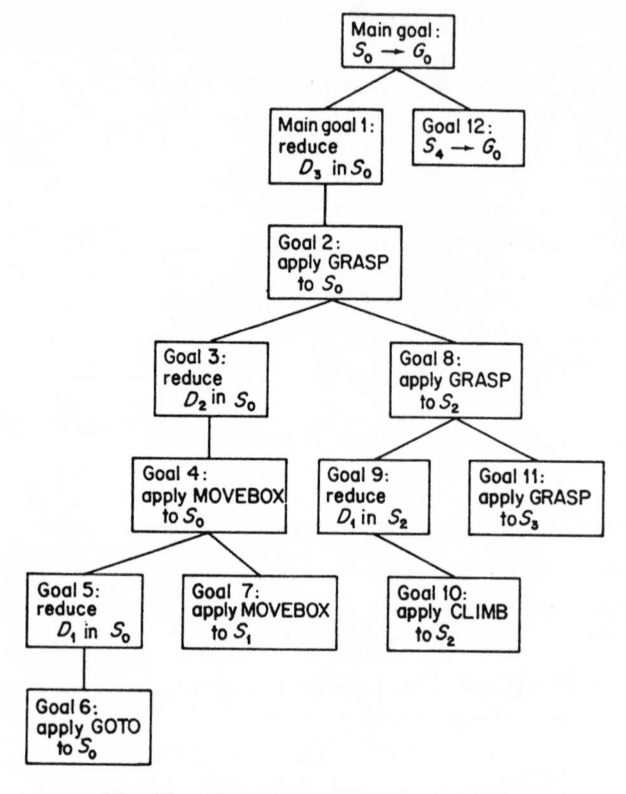

Fig. 38. Example of operation of GPS

5.2.2 Operation of GPS

The problem of changing from an initial state S_0 to a goal state G_0 is given to GPS. Specifically, as shown in Fig. 38, the main goal is $S_0 \rightarrow G_0$. GPS examines the differences between G_0 and S_0. If there is no difference the main goal is reached, and the problem is solved; if there still exist differences, it selects operators which can reduce them. In the problem of the monkey and banana, D_3 is a difference. Accordingly, in this figure, the subgoal of reducing D_3 is generated; that is, the subgoal is $S_0 \rightarrow$ HOLD(banana). There is only one operator which is effective for this subgoal, namely GRASP(banana). GPS, therefore, generates the subgoal of applying GRASP(banana) to S_0. For this to be done, the preconditions of GRASP must be satisfied. GPS examines the differences between the preconditions and S_0.

In this example the differences are D_1 and D_2. If there are several differences, reduction of the most important difference is given priority. Reduction of D_2 therefore becomes the subgoal. That is, the subgoal is $S_0 \rightarrow$ AT(box, c). The only operators related to D_2 are MOVEBOX(v). If v = c the subgoal is reached, so the next subgoal is to apply MOVEBOX(c) to S_0. The relationship between the preconditions and S_0 is then re-examined. The states related to the differences are:

S_0: AT(monkey, a) ∧ AT(box, b)
Precondition: AT(monkey, x) ∧ AT(box, x)

x is arbitrary; so if we let it be the most important item—namely b, the position of the box—the subgoal $S_0 \rightarrow$ AT(monkey, b) is obtained. This is reached by the operator GOTO(u), so we take u = b. Because GOTO(u) has no preconditions, GOTO(b) can be immediately applied to S_0. The state S_1 immediately after it has been applied is as follows:

S_1: AT(monkey, b) ∧ AT(box, b) ∧ EMPTY

Since the difference D_1 has now indeed been eliminated, we attempt to reach the immediately preceding subgoal (applying MOVEBOX(c)). Since the precondition is satisfied, application of MOVEBOX (c) to S_1 results in the new state S_2:

S_2: AT(monkey, c) ∧ AT(box, c) ∧ EMPTY

It will also be noted that D_2 has now been reduced. We therefore try to reach the immediately preceding subgoal, GRASP(banana). The precondition for this is not satisfied; so, by examining the difference between the precondition and S_2, we generate the new subgoal:

$S_2 \rightarrow$ ON(monkey, box)

The operator for movement of the monkey in the vertical direction is CLIMB, so this operator is selected. S_2 satisfies the precondition of CLIMB, so it can be immediately applied to S_2, resulting in S_3:

S_3: AT(box, c) ∧ ON(monkey, box) ∧ EMPTY

Since S_3 satisfies the precondition of GRASP, we apply GRASP to change S_3 to S_4:

S_4: AT(box, c) ∧ ON(monkey, box) ∧ HOLD(banana)

It will be noted that D_3 has now been reduced. We return to the immediately preceding goal. This is:

$S_4 \rightarrow G_0$

and we can see that S_4 has already satisfied G_0. Since this goal is the main goal, the problem is solved.

In this example, the operators to reduce the differences were always selected correctly. However, in general there are several candidate operators for reducing given differences some of which, if selected, might not be successful. If they are unsuccessful, another operator or difference should be selected.

GPS performs a depth-first search from the goal state to the initial state. However, the selection of differences and operators is not made at random, but is restricted by a table showing the relationship between the operators and differences, and the order of importance of the differences. This technique is called means–ends analysis.

5.2.3 GPS algorithm

To formalize the operation of GPS we shall introduce the concept of a *recursive procedure*. If a procedure uses itself in the definition, it is called a recursive procedure (For details, see Chapter 6). The GPS procedure that converts state S into state G can be defined as follows:

Recursive procedure *GPS* (S, G)

1. **If** $S \supset G$ **then** *return* (S); if the state S satisfies the state G then the goal is reached.
2. Find the differences between G and S. Put them in a list called *difference*, in order of importance.
3. *LOOP*1: **if** *empty* (*difference*) **then** *return* (*fail*).
4. d: = *first* (*difference*). *Remove* (d, *difference*).
5. Find all the operators that reduce d, and put them in a list called *operator*.
6. *LOOP*2: **if** *empty* (*operator*) **then** *goto LOOP*1.
7. *op*: = *first* (*operator*). *Remove* (*op*, *operator*).
8. *pc*: = precondition of *op*.
9. S_1: = *GPS* (S, *pc*); S_1 is the state after application of *GPS* (S, *pc*) or *fail*.
10. **If** S_1 = *fail* **then** *goto LOOP*2.
11. S_2: = *op* (S_1); apply *op* to S_1.
12. S_3: = *GPS* (S_2, G).
13. **If** S_3 = *fail* **then** *goto LOOP*2.
14. *Return* (S_3).

If the given problem is to change the initial state S_0 to the goal state G_0, this is solved by GPS(S_0, G_0). In the example of the monkey and banana, when we enter GPS(S_0, G_0), D_3 is chosen in step 4, and GRASP is chosen in step 7. The precondition *pc* of GRASP is AT(box, c) ∧ ON(monkey, box). GPS therefore divides the goal $S \rightarrow G$ into the sub-goal of changing from S to the state S_1 satisfying *pc*, the sub-goal of applying GRASP to S_1, and the sub-goal of changing this result S_2 to the state S_3 satisfying G_0. The above three sub-goals are reached by step 9, step 11, and step 12 respectively. Each of these sub-goals is simpler than the main goal. Thus, when GPS is used recursively, the sub-goals get progressively simpler, so we may expect that they will eventually be attained. For example, in step 7 of the procedure GPS(S, G) for attaining the precondition of MOVEBOX(b, c), *op*: = GOTO(b) is selected. GOTO has no preconditions, so GPS(S, NIL) of step 9 is immediately successful. (NIL means that there are none.) In step 11, S_2 becomes S_1 of Fig. 38. Since S_2 satisfies G, GPS(S_2, G) of step 12 is immediately successful, and GPS(S, G) terminates.

It should be noted that, in this procedure, the sequence of application of the operators is not stored. If this is needed, a list of operators that are applied can be stored as the procedure runs. That is, we can arrange so that the result given by GPS includes not just the obtained state, but also the sequence of operators which was applied. If the operator sequence obtained in step 9 is L_1, and the operator sequence obtained in step 12 is L_2, the operator sequence L of this GPS(S, G) is obtained by inserting *op* between L_1 and L_2. Step 14 should give S_3 and L.

5.3 PROBLEM-SOLVING BY A RESOLUTION PRINCIPLE

We have now described various methods of representing a problem by a state space, and solving it by searching this state space. These may be handled in a uniform manner by logic. The problem is represented by predicate calculus, and solved using logical inference. Considerable research has gone into predicate calculus and methods of inference.‡ Here we shall describe the first-order predicate calculus, which is widely used for solving problems, and the resolution principle, which is a typical method of deductive inference.

5.3.1 First-order predicate calculus

Predicate calculus is a formal language used to represent various facts. First-order predicate calculus is more restricted than other (higher-order) forms, but is capable of representing most mathematical facts and the facts which we deal with in our everyday lives.

In first-order predicate calculus, facts are represented using constants, variables, and functions which are used in ordinary mathematics, and, in addition, predicates. For example, in the previous section the situation that the monkey was on the box was expressed by the formula:

ON(monkey, box)

In this formula, ON is a predicate symbol and 'monkey' and 'box' are constant symbols representing specific objects. The above formula is called an *atomic formula*. If the positions of the monkey and box are given, the formula has the

‡ For details, see No. 7 in the Iwanami Information Science Series *Logic and Meaning*.

truth value either true (T) or false (F). When x and y are both elements of the set D, ON(x, y) is an atomic formula. The set D is called the domain, and x and y are variable symbols. If we represent the child of x by the function *child*(x), and the object closest to y by the function *neighbour*(y), *child* and *neighbour* are function symbols. Atomic formulae may include functions. For example:

ON(child(x), neighbour(y))

In general, an atomic formula consists of a predicate symbol and a term. A term is a constant symbol, variable symbol or function expression. T or F themselves are atomic formulae. We can write the fact that x and y are equal as the atomic formula EQUAL(x, y); but we can also write it directly as $(x = y)$. A first-order predicate calculus that permits such an expression may be called a first-order predicate calculus with equality.

In this formal predicate calculus language, well-formed formulae (abbreviated to wff) correspond to statements of the language. The simplest wff are atomic formulae. If P and Q are wff, then $(\sim P)$, $(P \wedge Q)$, $(P \vee Q)$, and $(P \equiv Q)$ $(P \Rightarrow Q)$ are also wff. $\sim$, $\wedge$, $\vee$, $\equiv$, $\Rightarrow$ are respectively the symbols for NOT, AND, OR, *equivalence*, and *implication*. They are called *logical connectives*. It should be noted that '='—a predicate—is used to connect terms and '≡'—a logical connective—is used to connect wff. If $Q \equiv (\sim P)$, the truth values of P and Q are different. Apart from $\sim$, all of these connect two wff. The relationships between two wff and connected wff are shown by the truth table shown as Table 10. As is clear, $P \Rightarrow Q$ and $\sim P \vee Q$ are equivalent.

TABLE 10
Truth table of wff including connected symbols

P	Q	$P \wedge Q$	$P \vee Q$	$P \Rightarrow Q$	$P \equiv Q$
T	T	T	T	T	T
T	F	F	T	F	F
F	T	F	T	T	F
F	F	F	F	T	T

If we let x be a real number, and x^2 be $f(x)$, and represent $x \leqslant y$ as LESS(x, y), that $0 \leqslant x^2$ for all x can be expressed as follows:

$(\forall x)$LESS$(0, f(x))$

The wff that, for any x, there exists a real number y that is not less than x may be expressed as follows:

$(\forall x)$ [$(\exists y)$LESS(x, y)]

$\forall$ is the universal quantifier and $\exists$ is the existence quantifier. The above formula can be written as $(\forall x)\,(\exists y)\mathrm{LESS}(x, y)$ without affecting the meaning. The following wff is also true:

$$(\forall x)\,(\forall y)\,(\forall z)\,[\{\mathrm{LESS}(x, y) \wedge \mathrm{LESS}(y, z)\} \Rightarrow \mathrm{LESS}(x, z)]$$

The truth value of the wff given above is determined when suitable interpretations are given to the predicate symbols or function symbols. However, there are some wff whose truth value does not change no matter what interpretation is given. A wff which is true for any interpretation is called a *valid wff*, and a wff which is false for any interpretation is called an *unsatisfiable wff*. Examples of these now be given. In the following, a, b, and c are constant symbols, and u and x, etc. are variable symbols.

$$(\forall x)\,P(x) \Rightarrow P(a)$$
$$\sim P(a) \wedge P(a)$$

5.3.2 Clause forms

An atomic formula or the negation of an atomic formula is called a *literal*. A literal or the disjunction of literals is called a *clause*. For example, $P(x) \vee \sim Q(x, f(x)) \vee R(x, y)$ is a clause. A wff expressed in the form:

$$(\forall x_1)\,(\forall x_2)\,\ldots\,(\forall x_n)\,[C_1 \wedge C_2 \wedge \ldots \wedge C_m],$$

where C_i $(1 \leqslant i \leqslant m)$ are clauses and x_1, x_2, ..., x_n are all the variables appearing in the clauses, is called a *clause-form wff*. The part within the square brackets in the above formulas is called the *matrix*. The following steps turn a given wff into clause form:

1. Change $P \Rightarrow Q$ into $\sim P \vee Q$, and $A \equiv B$ into $(\sim A \vee B) \wedge (\sim B \vee A)$.
2. Standardize the variables. For example, in $(\forall x)\,[\sim P(x) \vee (\exists x)Q(x)]$ the x that appear in $(\forall x)$ and $(\exists x)$ represent different variables, so we write this as $(\forall x)\,[\sim P(x) \vee (\exists y)Q(y)]$.
3. Restrict the range of negation. For example, change $\sim (\forall x)\,[P \wedge \sim (Q \wedge R)]$ into $(\exists x)\,[\sim P \vee (Q \wedge R)]$.
4. Eliminate existential quantifiers. For example, $(\forall x)\,(\exists y)\,P(x, y)$ means that for any x, if x is determined, there exists a y that satisfies $P(x, y)$. y can therefore be regarded as a function of x. We may therefore convert this to $(\forall x)\,P(x, f(x))$. This is called a *Skolem function*.
5. Bring the universal quantifiers to the front. For example, convert $(\forall x)\,[P(x) \vee (\forall y) \sim Q(y)]$ into $(\forall x)\,(\forall y)\,[P(x) \vee \sim Q(y)]$.
6. Put the matrix in conjunctive normal form. $\wedge$ should be distributed with respect to $\vee$. For example, $(A \wedge B) \vee (B \wedge C)$ should be written as $(A \vee B) \wedge (B \vee C) \wedge B \wedge (A \vee C)$ (for simplicity).

A universal quantifier will always appear in front of the function that appears in the matrix of a clause-form wff, so this can be omitted. Clauses will always be connected up by $\wedge$, so we only need to list the clauses. For example:

$$(\forall x_1)(\forall x_2)[P(x_1, f(x_2)) \wedge (P(g(x_1), x_2) \vee Q(x_1, x_2))]$$

becomes:

$$P(x_1, f(x_2)) \quad (8)$$

and

$$P(g(x_1), x_2) \vee Q(x_1, x_2) \quad (9)$$

Note that the x_1 and x_2 are universally quantified, so we do not always need to use the same variable names. In fact, it is more convenient, so far as application of the resolution principle is concerned, if we use different variable names. We therefore transform clause (9) as follows:

$$P(g(x_3), x_4) \vee Q(x_3, x_4) \quad (9')$$

5.3.3 Resolution principle

Consider the problem of proving that a wff expressed in clause form is unsatisfiable.

Assume that one clause contains a literal P, and another clause includes the negative $\sim P$ of P (P and $\sim P$ are said to be *complementary*). Specifically, the clauses are:

$$P \vee Q_1 \vee Q_2 \vee \ldots \vee Q_m \quad (10)$$
$$\sim P \vee R_1 \vee R_2 \vee \ldots \vee R_n \quad (11)$$

Clauses (10) and (11) are respectively equivalent to $\sim P \Rightarrow \{Q_1 \vee Q_2 \vee \ldots \vee Q_m\}$ and $P \Rightarrow \{R_1 \vee R_2 \vee \ldots \vee R_n\}$. Since P is either true or false, the following clause can be derived:

$$Q_1 \vee Q_2 \vee \ldots \vee Q_m \vee R_1 \vee R_2 \vee \ldots \vee R_n \quad (12)$$

Using two clauses C_1 and C_2 containing complementary literals P and $\sim P$ to produce a disjunction consisting of the literals contained in C_1 and C_2 apart from P and $\sim P$ is called *resolution*. The resultant clause is called the *resolvent* clause, and C_1 and C_2 are called *parent* clauses. If the original set of clause forms cannot be satisfied the set of clause forms with the addition of the resolvent clause is also unsatisfiable. If a clause form contains a single literal clause and the complementary single literal clause, it is clearly unsatisfiable. If these complementary clauses are represented by P and $\sim P$, the empty clause (written NIL) can be derived from them.

For example, we can prove that the following clause forms:

$$P(a) \vee Q(b) \quad (13)$$

$$\sim P(a) \tag{14}$$
$$\sim Q(b) \tag{15}$$

are unsatisfiable by deriving $Q(b)$ from clauses (13) and (14) and deriving the empty clause from $Q(b)$ and $\sim Q(b)$ from clause (15).

A proof based on the principle of resolution shows that the negation of the wff which is to be proved is unsatisfiable. If the wff which is to be proved is written as $P \Rightarrow Q$, its negation is $P \wedge \sim Q$. P is called the *axiom*, Q the *goal* wff, and the clause form $\sim Q$ the *negation* of the goal wff.
For example, if the negation of the wff:

$$[(\forall x)(\exists v)P(x, v) \wedge (\forall y)(\forall z)[P(a, y) \Rightarrow \sim Q(y, z)]]$$
$$\Rightarrow \sim(\forall u)(\exists w)Q(u, w)$$

is expressed as clause forms, we have:

$$P(x, f(x)) \tag{16}$$
$$\sim P(a, y) \vee \sim Q(y, z) \tag{17}$$
$$Q(u, g(u)) \tag{18}$$

Clauses (16)–(18) are the set constituting the starting points of the proof, so they are called the *input* clauses. We cannot immediately generate the resolvent clause from these input clauses. However, if we substitute a for x in clause (16), and substitute $f(a)$ for y in clause (17), we obtain:

$$P(a, f(a)) \tag{16$'$}$$
$$\sim P(a, f(a)) \vee \sim Q(f(a), z) \tag{17$'$}$$

From these, the following clause is derived:

$$\sim Q(f(a), z) \tag{19}$$

Similarly, the empty clause is deduced by substituting $f(a)$ for u in clause (18), and $g(f(a))$ for z in clause (19). Substituting suitable terms for variables contained in literals, so that the resolvent clause can be derived, is called *unification*. For theorem-proving using the principle of resolution to be carried out automatically, a method of unification and a method of controlling problem-solving to determine which clauses are to serve as the parents for the resolvent clauses are needed.

5.3.4 Method of unification

Unification means the insertion of suitable terms as the variables contained in several literals to make all the literals the same. Substituting terms t_i for variables v_i is written as $(v_i \leftarrow t_i)$. This substitution is written as s_i. For example, if we take $s_1 = y \leftarrow a$:

$$P(y, z)s_1 = P(a, z)$$

Carrying out a substitution s_2 after performing a substitution s_1 is written as $s_1 s_2$. For example, if $s_2 = (z \leftarrow f(x))$:

$$P(y, z)s_1 s_2 = P(a, f(x))$$

The two literals $\{P(y, f(x)), P(a, z)\}$ can both be converted into $P(a, f(x))$ by $s_1 s_2$. Such a substitution $s = s_1 s_2$ is called a *unifier*. If we take $s_3 = (x \leftarrow b)$, $s' = s_1 s_2 s_3$ unifies the same two literals to $P(a, f(b))$. s' is also a unifier. However, s' is the result of applying the substitution s_3 to s. A unifier such as s that can derive all other unifiers by applying a suitable substitution is called a *most general* unifier. To obtain the resolvent clause, we have to find the most general unifier.

Let us consider a general method of obtaining a most general unifier of $\{P(f(x), x, g(x))P(y, a, x)\}$. First, if two literals contain the same variable, change the name of one variable. In this example, x is contained in both of the literals, so we change the x that appears in the latter to z, writing $P(y, a, z)$. Next, regarding the two literals as respective series of symbols, examine places of disagreement of the symbols, starting from the left. We find that f and y are in disagreement. The term starting with f is $f(x)$, and the term starting with y is y. The disagreement term $\{f(x), y\}$ is called the *disagreement set*. The element y of the disagreement set is a variable. So we take substitution s_1 as $(y \leftarrow f(x))$. If we apply s_1, the two literals are converted to $\{P(f(x), x, g(x)), P(f(x), a, z)\}$. If we again look for the disagreement set, we obtain $\{x, a\}$, so we find $s_2 = (x \leftarrow a)$. Carrying out s_2, we obtain $\{P(f(a), a, g(a)), P(f(a), a, z)\}$.

Next, from the disagreement set $\{g(a), z\}$, we find $s_3 = (z \leftarrow g(a))$, so by s_3, the literals are unified to $P(f(a), a, g, (a))$. In other words, the required most general unifier is $s = s_1 s_2 s_3$.

An algorithm to find the most general unifier if two literals $\{P_1$ and $P_2\}$ do not contain the same variables may be expressed by the following procedure:

Procedure *unify* (P_1, P_2)

1. $s :=$ empty substitution (the empty substitution does not change anything). $Q_1 := P_1$, $Q_2 := P_2$
2. *LOOP*: find the disagreement set D of Q_1 and Q_2.
3. **If** *empty*(D) **then** *return*(s).
4. $t_1 = first(D)$, $t_2 = last(D)$.
5. **If** *variable*(t_1) and *not-contain*(t_2, t_1) **then** $s_1 := (t_1 \leftarrow t_2)$.
   else if *variable*(t_2) and *not-contain*(t_1, t_2) **then** $s_1 := (t_2 \leftarrow t_1)$
   else *return*(*fail*).
6. $Q_1 := Q_1 s_1$.
   $Q_2 := Q_2 s_1$.
   $s = s s_1$.
7. **Goto** *LOOP*.

In this procedure, if we express the disagreement set D as $D = \{t_1, t_2\}$ *first*(D) is t_1 and last(D) is t_2. *Variable*(t) is a predicate function that is true if t is a variable, but otherwise false; and *not-contain*(x, y) is a predicate function that is true if t_2 does not contain t_1. For example, x and $f(x)$ cannot be unified.

5.3.5 Control of proof based on principle of resolution

Control of problem-solving can be performed by a strategy of selecting the parent clauses in order to derive the empty clause by the principle of resolution. A typical strategy is shown below. For convenience in explanation we shall use the following example.

Example

If:

$$[(\forall u)(\forall v)\{R(u) \Rightarrow P(u, v)\}] \wedge [(\forall w)(\forall z)\{Q(w, z) \Rightarrow S(z)\}] \wedge R(b) \wedge \sim S(b)$$

is true, then prove $(\exists x)(\exists y)\{P(x, y) \wedge \sim Q(x, y)\}$. For this proof, we need to show that the following clause forms, which are negations of *wff*, are unsatisfiable:

$$\sim P(x,y) \vee Q(x,y) \quad (20)$$
$$\sim R(u) \vee P(u,v) \quad (21)$$
$$\sim Q(w,z) \vee S(z) \quad (22)$$
$$R(b) \quad (23)$$
$$\sim S(b) \quad (24)$$

Clause (20) is the negative of the goal wff. The rest are axioms.

(i) *Breadth-first strategy* Initially we find all the resolvent clauses that have input clauses as parents. Next, we find all the resolvent clauses which have as parents input clauses and the resolvent clause which have been obtained. This process of finding resolvent clauses at successively deeper levels is continued until the empty clauses is obtained.

Fig. 39 shows how resolution proceeds as the breadth-first strategy is applied to the example (unification is omitted). In some cases the same clauses are obtained as resolvent clauses of depth 2. It is, of course, unnecessary to add such resolvent clauses again to the clause form when they are obtained for the second time. The breadth-first strategy can derive the empty clause at the shallowest level, but it has the drawback that in general it generates a large number of resolvent clauses.

(ii) *Linear resolution* In linear resolution, when resolvent clauses are generated, initially they have two input clauses; but when this process is carried out for the second and subsequent times, one parent is always the immediately

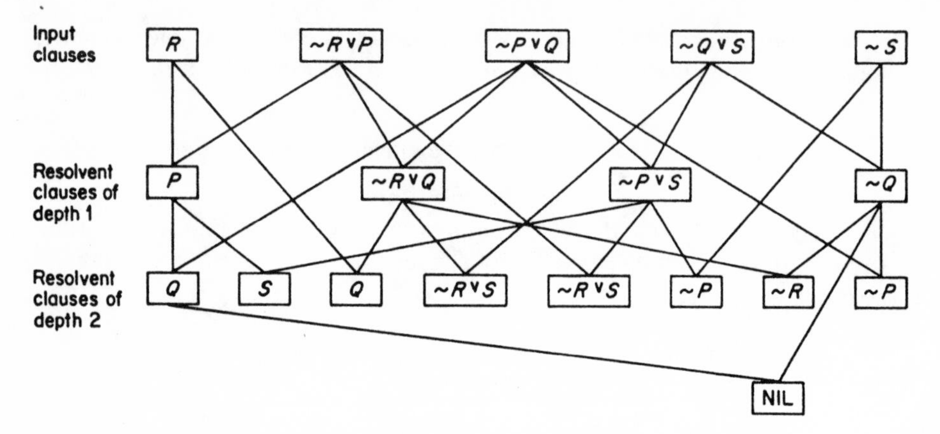

Fig. 39. Process of resolution by breadth-first strategy

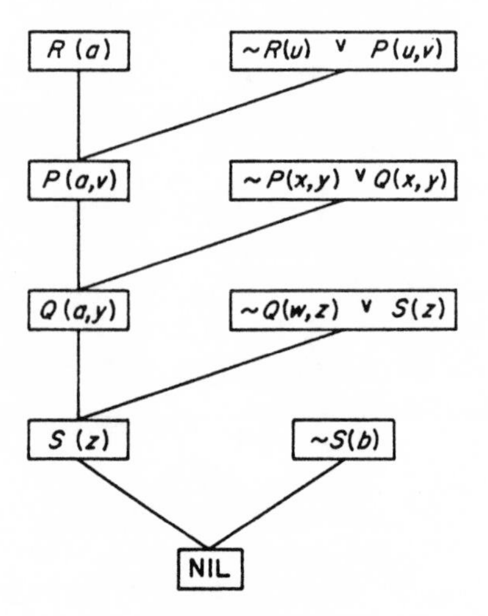

Fig. 40. Resolution process using linear resolution

preceding resolvent clause. An example is shown in Fig. 40. It can be proved that, if the set of input nodes is unsatisfiable this strategy will always lead to a contradiction. Thus the linear resolution is said to be complete.

(iii) *Set of support strategy* If the set of clauses S is unsatisfiable, and the set $S - T$ of the remaining clauses after the set of clauses T has been taken from S is satisfiable, T is called a *set of support*.

Assume that the wff to be proved in $P \Rightarrow Q$, and that P is satisfiable. Thus the set of input clauses using the resolution principle are clause forms on P and $\sim Q$. That is, the negation of the goal wff is a support set, corresponding to the

T just referred to. The 'set of support strategy' consists in, when resolving, selecting at least one parent from a set of support or a clause resolved from a set of support. In the example, the set of support is $\sim P(x, y) \vee Q(x, y)$, and the empty clause is resolved at the level of depth 4 (the reader should verify this). This shows that the set of support strategy is also complete.

(iv) *Unit preference strategy* The unit preference strategy is to give priority to clauses consisting of only one literal, and attempt to derive resolvent clauses which have these as parents. Since the resolved clauses in their turn will only contain a few literals, the empty clause is likely to be resolved sooner. Fig. 40 also exemplifies the unit preference strategy. Of course, if unit clauses cannot be selected, the resolution must be continued by some other method.

(v) *Other strategies* In another strategy, at least one of the parents of the resolvent clause is selected from the input clauses. There is no guarantee that such a strategy will always find the solution. Another strategy, which will always find a solution, is to select at least one of the parents of a resolvent clause, either from the input clauses or from the ancestor of another parent.

It is also possible for several strategies to be combined. For example, in the unit preference strategy, if there are several candidate clauses, the clauses of the set of support can be given priority.

Another possibility is to improve the efficiency of the resolution process by constructing a graph in which all the literals appearing in a clause form constitute nodes of the graph and complementary literals are connected by edges.‡ Another method is to attempt unification only when a candidate solution has been found by resolution. But even these methods require a strategy for deciding which resolution step is to be performed first. They may therefore advantageously be combined with the various strategies described earlier.

5.3.6 Representation of the answer

As mentioned earlier, we can prove a wff by showing that its negation is unsatisfiable. Now let us consider how this answer should be represented to make it easier to understand.

The first possibility is to take the resolution process itself as the answer. Using the resolution process of Fig. 40, for example, it is easy to see that negating the goal wff leads to a contradiction. That is, since all the clauses except $\sim P(x, y) \vee Q(x, y)$ are axioms, $P(a, v)$, which is initially derived from $R(a)$ and $\sim R(u) \vee P(u, v)$, is a theorem. $Q(a, y)$ is derived from the negation of

‡ Sickel, S., 'A search technique for clause interconnectivity graphs', *IEEE Trans.*, **C25**, 1976, 823–835.

this theorem and the goal wff, and $S(z)$ is derived from this and the axiom $\sim Q(w, z) \vee S(z)$. This contradicts the axiom $\sim S(b)$.

The answer could also be expressed positively. In this method, the answer is obtained by the process of deriving the empty clause by combining literals with each clause of the negation of the goal wff to obtain a clause (tautology) which is necessarily true. Taking the same example as before, the negation of the goal wff is:

$$\sim P(x, y) \vee Q(x, y)$$

Combining $P(x, y) \wedge \sim Q(x, y)$ with this we construct the tautology:

$$\sim P(x, y) \vee Q(x, y) \vee (P(x, y) \wedge \sim Q(x, y))$$

If we use this to perform the resolution shown in Fig. 40, we obtain, as shown in Fig. 41:

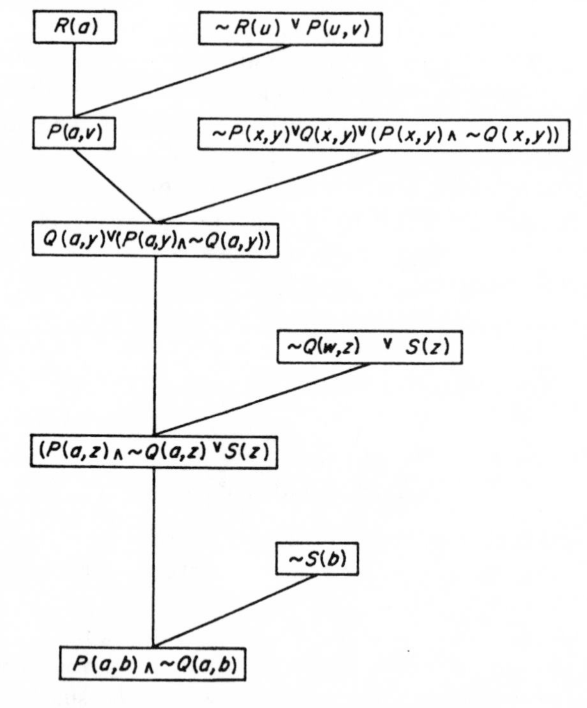

Fig. 41. Resolution using tautology

$$P(a, b) \wedge \sim Q(a, b)$$

This shows that the goal wff of the original problem $(\exists x)\,(\exists y)\,\{P(x, y) \wedge \sim Q(x, y)\}$ has the solution $x = a$, $y = b$.

However, the answer will not necessarily be as simple as this. Sometimes the wff finally obtained will contain a skolem function. In such cases, the wff must be interpreted taking into consideration the meaning of the skolem function.

5.4 PLANS

When the procedure for solving a given problem becomes fairly complicated, a powerful strategy is to establish a rather general plan for obtaining the solution. As mentioned earlier, GPS is based on the plan of reducing the most important differences. This section is concerned with how plans should be formulated to enable a robot to solve various problems. We shall first describe methods of representing problems and methods of searching to obtain solutions, then how they can be improved by use of a plan.

5.4.1 Representation of problems and basic methods of solution

Solving a problem involves converting a given initial state into a goal state. States can be represented by the predicate calculus. The operators are given and there is a definition, for each operator, of how the state is changed by the application of the preconditions and the operator.

For example, the problem shown in Fig. 42 may be represented as follows:

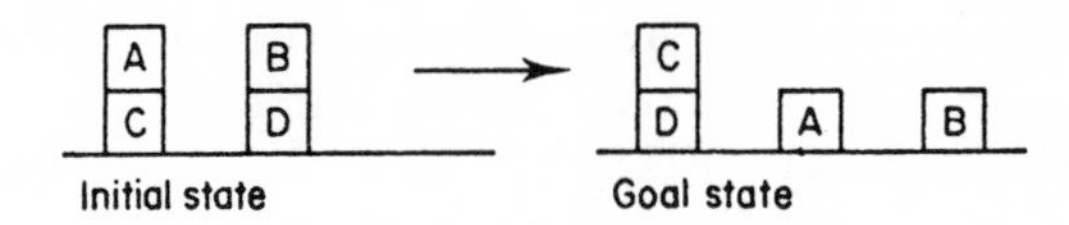

Fig. 42. Simple blocks problem

(i) *Initial state*
EMPTY; the robot's hand is not grasping anything
ONTABLE(C)
ON(A, C)
ONTABLE(D)
ON(B, D)

(ii) *Goal state*
ON(C, D); nothing is specified regarding A and B, so they need not be in exactly the positions shown in the goal state of Fig. 42.

(iii) *Operators*
PICKUP(x); pick up the x on the table.
  preconditions: ONTABLE(x), CLEAR(x), EMPTY
  delete list: ONTABLE(x), CLEAR(x), EMPTY
  add list: HOLD(x), the robot's hand is holding x.

PUTDOWN(x); put x down on the table.
precondition: HOLD(x)
delete list: HOLD(x)
add list: ONTABLE(x), CLEAR(x), EMPTY
TAKEOFF(x, y): grasp the x that is on top of y.
preconditions: ON(x, y), CLEAR(x), EMPTY
delete list: ON(x, y), CLEAR(x), EMPTY
add list: CLEAR(y), HOLD(x).
PUTON(x, y); put x on top of y.
preconditions: HOLD(x), CLEAR(y)
delete list: HOLD(x), CLEAR(y)
add list: ON(x, y), CLEAR(x), EMPTY.

This problem can be solved as follows. Express the goal state wff in clause form and find the clauses that cannot be derived from the initial state. Make a suitable choice of one clause C_i from among these clauses. Find an operator that has C_i in the add list. Check whether its preconditions are satisfied by the initial state. If they are satisfied, the clause C_i in question is satisfied by application of this operator. If they are not satisfied, take as a sub-goal the clause form in which the clause C_i is replaced by the preconditions of the operator. By repeating this process, eventually all the clauses will be satisfied by the initial state and the problem solved. If more than one operator can be applied to achieve the sub-goal, this gives rise to a decision point in the search.

In the example of Fig. 42, the goal ON(C, D) is not satisfied by the initial state, so we select an operator PUTON(x, y) that contains ON(x, y) in its add list. This procedure can be performed using the resolution principle. That is, if the empty clause can be derived from the negation ~ON(C, B) of the goal and the add list of the operator, we know that that operator can be selected. In this example, the empty clause is derived by the substitution ($x \leftarrow$ C) ($y \leftarrow$ D). The sub-goal is therefore the preconditions of PUTON(C, D), that is HOLD(C), CLEAR(D). Next look at either of the clauses HOLD(C) and CLEAR(D) and select an operator. Where the sub-goal consists of more than one clause, even if an operator satisfies one of these clauses, it is not to be selected if the other clause is inconsistent with the operator's delete list or add list. For example, the operator PUTDOWN(D) which has CLEAR(D) in its add list also contains EMPTY in its add list. This is inconsistent with the other goal, HOLD(C), so this operator is not selected. The first half of the solution process of this example is shown in Fig. 43. The constraints on x are applied in state (3). Comparing states (2) and (3), (2) has fewer clauses which are not satisfied by the initial state, so (2) is expanded. The clauses in (2) which are not satisfied by the initial state are CLEAR(C) and CLEAR(D). Either of these may be given priority. The diagram shows only the case where the former is given priority. The clauses in state (6) apart from CLEAR(D) are satisfied by the

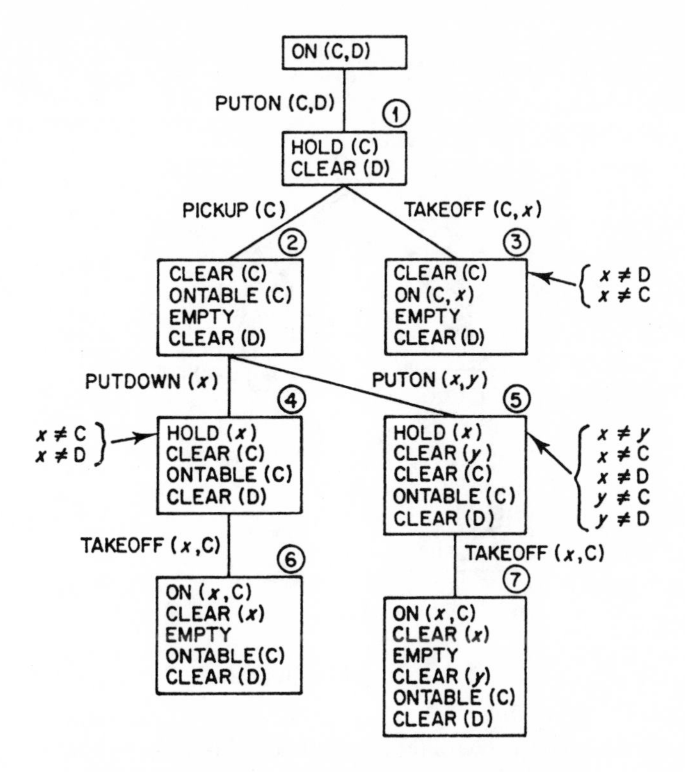

Fig. 43. First half of solution process of Fig. 42

initial state if we make the substitution $(x \leftarrow \mathrm{A})$, so $(x \leftarrow \mathrm{A})$ is performed on the clauses of (6), the clauses of (4), and the intervening operators. The variable constraints are shown to the right of state (5). If, for example, we put $y = \mathrm{D}$, this would be contradictory to CLEAR(D) in (5). The substitution $(x \leftarrow \mathrm{A})$ is found in state (7). This substitution extends to (5), so $(y \leftarrow \mathrm{B})$ is obtained. Both (6) and (7) are on paths to a solution. If we follow the path through (7), the final state is ON(C, D), ON(A, B). This is also a solution. For simplicity, we only show the continuation from state (6) in Fig. 44.

It can be seen that state (8) is unrealizable. State (9) is satisfied by the initial state if we make the substitution $(x \leftarrow \mathrm{B})$. The solution is therefore:

$$\text{TAKEOFF(B, D)} \rightarrow \text{PUTDOWN(B)} \rightarrow \text{TAKEOFF(A, C)} \\ \rightarrow \text{PUTDOWN(A)} \rightarrow \text{PICKUP(C)} \rightarrow \text{PUTON(C, D)}$$

It will be noted that even a problem as simple as this example requires a certain amount of searching. Some complicated problems cannot be solved by the elementary method given here.

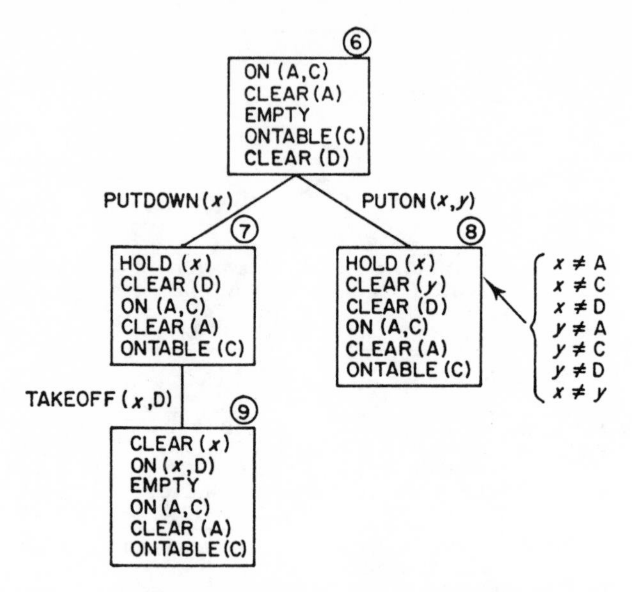

Fig. 44. Continuation from state (6)

5.4.2 Use of experience

Once a problem has been solved, it is convenient to remember the solution so that it can be used to solve similar problems. Alternatively, if, after a robot has found a sequence of mechanical operations, it encounters an unforeseen state when it puts this sequence of operations into practice, it may be necessary to change the plan. It has therefore been proposed to store the solution to a problem in a fixed form.

This fixed form is called the *triangle table*, and is shown in Fig. 45. The initial state is entered at row 0, column 0. The add list obtained by applying the operator O_1 to this is entered in row 1, column 1. In row 1, column 0, we enter the result of removing the delete list of O_1 from the initial state. In other words, the state after applying O_1 is shown in the columns of row 1. The result of applying O_2 to this state is shown in row 2. The state in the mth row after applying the last operator O_m satisfies the goal state.

Fig. 46 shows the triangle table for the solution of the example given above as far as the third row and third column. Next, for all operators, we put a mark against any clause in the row that is a precondition of an operator. In the last row, we put marks only against those clauses that are contained in the goal state. In this way we can see that the clauses that have been marked are the necessary conditions for obtaining a solution. This table contains constants, but the representation can be generalized by changing the constants into variables. Fig. 47 shows a generalized triangle table for this example. This figure shows only the marked clauses.

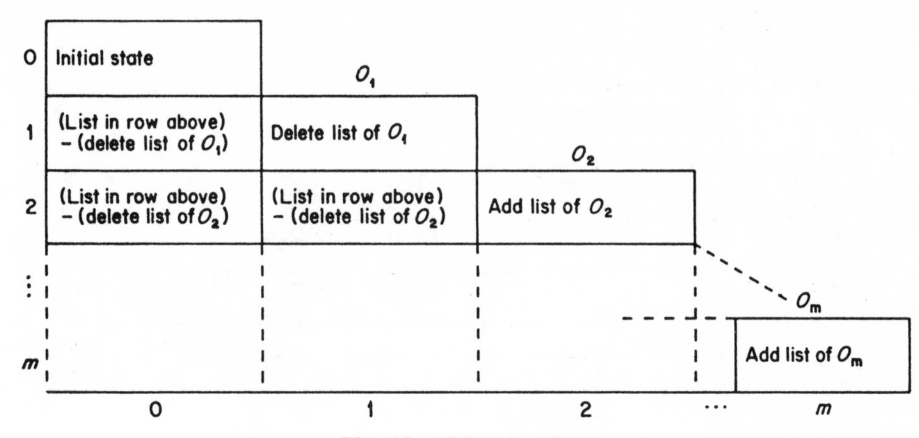

Fig. 45. Triangle table

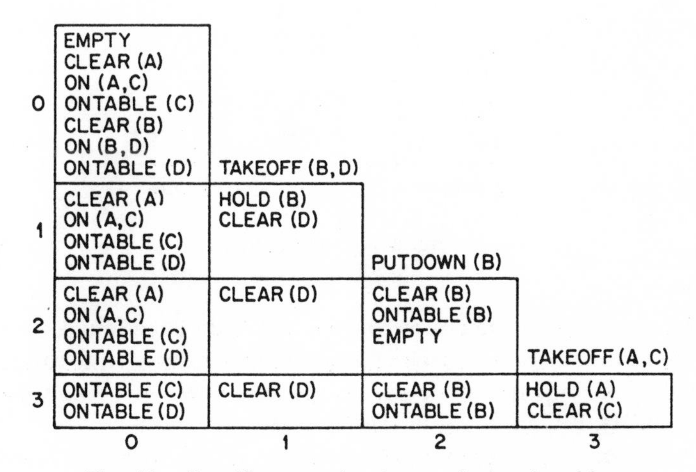

Fig. 46. Specific example of part of triangle table

If we look at the portion of the triangle table below the ith row and to the left of the ith column, we can see that the marked clauses in this portion are the preconditions for applying the operators which are found at numbers $i + 1$ to the end in the sequence of operators. For example, the portion below and to the left of the double line in Fig. 47 contains the five clauses that are the conditions for applying the third and following operators of the operator sequence. So any state that satisfies those clauses can be converted into the goal state.

Consider for instance a similar problem to that of the example described above. The goal ON(C, D) is exactly the same as before, but the initial state is a little different, as shown in Fig. 48. We can see that, in the triangle table of Fig. 47, if the substitutions ($u \leftarrow$ C) ($v \leftarrow$ D) ($x \leftarrow$ B) are carried out, the portion

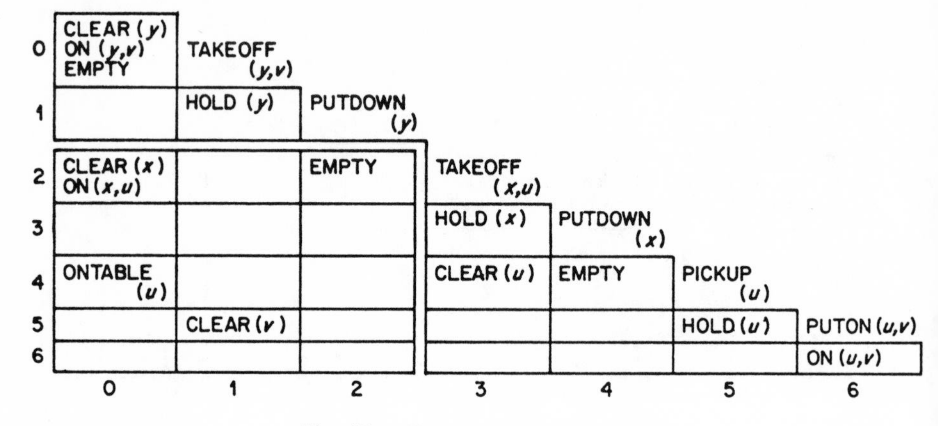

Fig. 47. Generalized triangle table

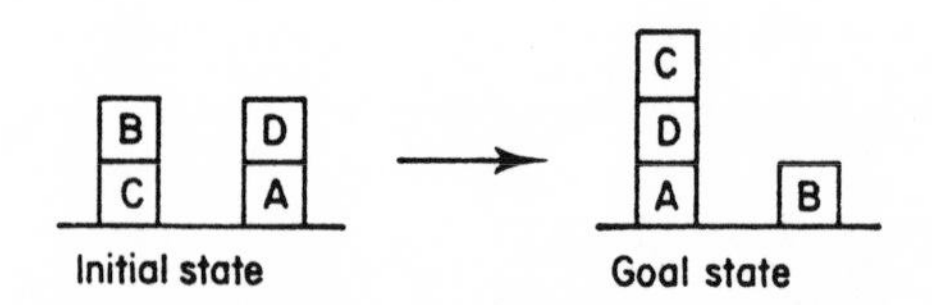

Fig. 48. Similar problem

below and to the left of the double lines is satisfied by the initial state, so the state of the sixth row coincides with the goal state. The goal state can therefore be reached by the sequence of operators from O_3 onwards: TAKEOFF(B, C), PUTDOWN(B), PICKUP(C), PUTON(C, D).

Now consider the situation which arises if the robot encounters an unforeseen state while executing the operating sequence shown in Fig. 47. For example, while executing the last operator PUTON(C, D), it drops the block C. Assuming that the robot can find C by using its eye, its state then satisfies the clauses in and below row 4 and to the left of column 4. So it can start again from PICKUP(C).

If we have triangle tables for several operations, we can move from states in one triangle table to states in another triangle table. If the clauses in the ith row of a triangle table (including unmarked clauses) satisfy the clauses below the ith row and to the left of the jth column of another triangle table, after executing operator O_i we can move to the $(j + 1)$th and subsequent operators of another triangle table.

Thus, if comparatively similar operations are carried out many times, it is useful to store the sequence of a typical operation in the form of a triangle table.

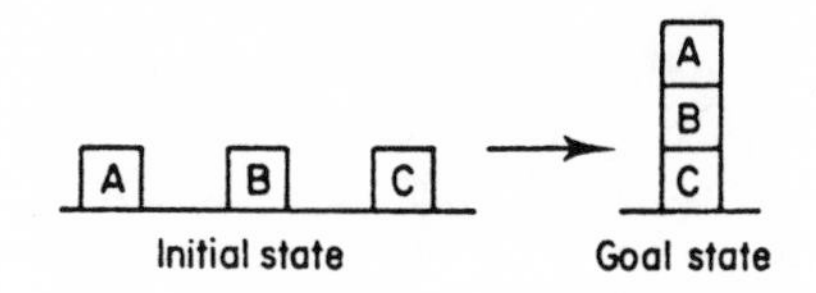

Fig. 49. Example of building a tower

5.4.3 Interference of several goals

In Section 5.4.1 we dealt with the problem of reaching a single goal; but during the process of obtaining a solution we had to attain sub-goals including several clauses. If these clauses are independent, solution of the problem can be simplified by solving these clauses separately. If they are not independent, we have to find a sequence of operations that attains all the goals at the same time. We shall now explain a method whereby several goals (clauses) can be solved independently, and then the complete solution obtained by considering the effect of mutual interference between these goals.

Consider the problem of Fig. 49 as a simple example. The operators are the same as before. The problem is broken down into several clauses, which are then solved, and represented by the AND/OR tree (there are no OR clauses) shown in Fig. 50. The operators are designated by arbitrary numbers. (Breaking down an initial goal into two goals is also considered as an operator.) Two operator sets are defined for each goal. The *add* operators of a goal C_i are the set of operators in the AND/OR tree other than the ancestors of C_i and

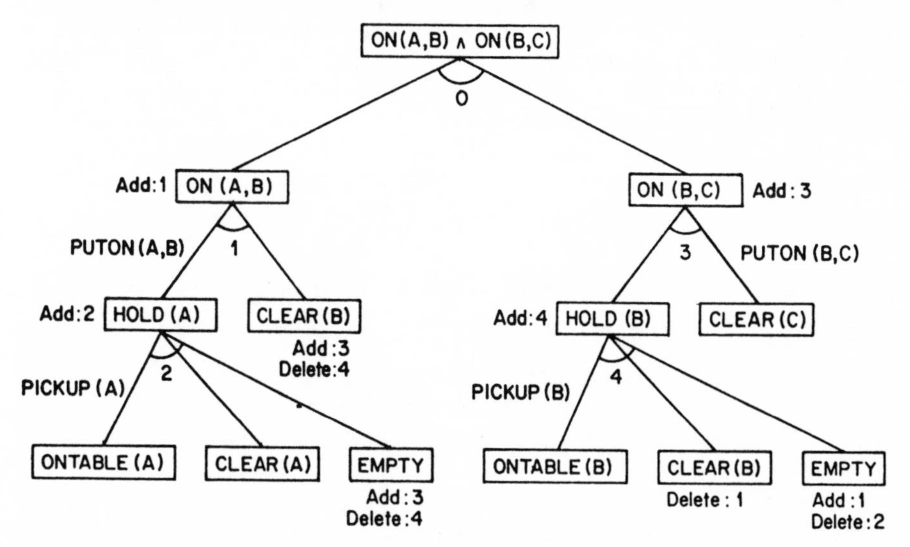

Fig. 50. AND/OR tree of a solution in which several goals are considered as independent

which contain C_i in their add list. The *delete* operators of a goal C_i are the set of operators in the AND/OR tree other than the ancestors of C_i and which contain C_i in their delete list. Both these sets are shown in Fig. 50 (in this case, the number of elements in each set is 1 or 0). For example CLEAR(B) is added by the operator PUTON(B, C) and is deleted by the operator HOLD(B), these operators not being ancestors of CLEAR(B).

The constraints of the operating sequence may be obtained from this figure. For any operator O_i to be applied, either the operators preceding it in the operator sequence must not be included in either the add list or the delete list of the preconditions of O_i, or the last operator in the sequence must be in the add list. For example, the precondition CLEAR(B) of the fourth operator in the figure includes 1 in its delete list, but not in its add list. So operator 4 cannot be applied after operator 1. If we examine the precondition EMPTY of operator 2, we can see that it cannot be applied immediately after 4, but would require for example the operator 3 to be inserted after the operator 4. The operator sequence 4, 3, 2, 1 is therefore a candidate solution. It is a complete solution since it satisfies the applicability conditions for all the operators.

5.4.4 Change of plan

Sometimes problems cannot be solved just by breaking them down and resynthesizing them. In such cases, the plan must be altered by adding or deleting operators.

An instance of this is provided by part of the example of Section 5.4.1. Assume that state (2) described in (a) has been attained. ONTABLE(C) is already satisfied by the initial state. The sub-goal is now, therefore, to achieve CLEAR(C) and EMPTY simultaneously. If we break this problem down as described in Section 5.4.3, we obtain Fig. 51.

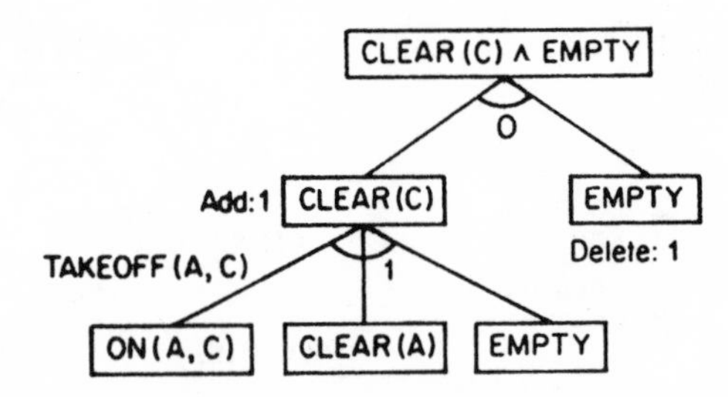

Fig. 51. AND/OR tree generated by the process of breaking down a problem

This sub-goal cannot be achieved directly since TAKEOFF(A, C), which is in the add list of CLEAR(C), is a deletion operator for EMPTY. We must therefore add an operator which has EMPTY in its add list. Since the add list of TAKEOFF(A, C) includes HOLD(A), we look for an operator that converts

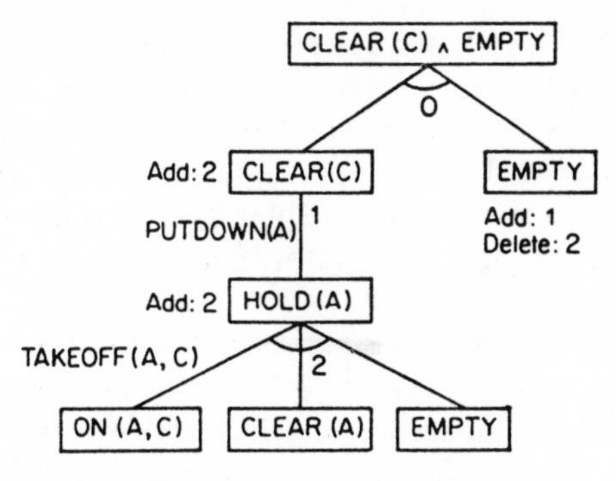

Fig. 52. AND/OR tree after insertion of an operator

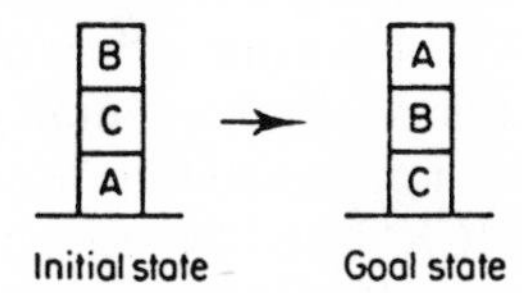

Fig. 53. Problem requiring extensive revision of the plan

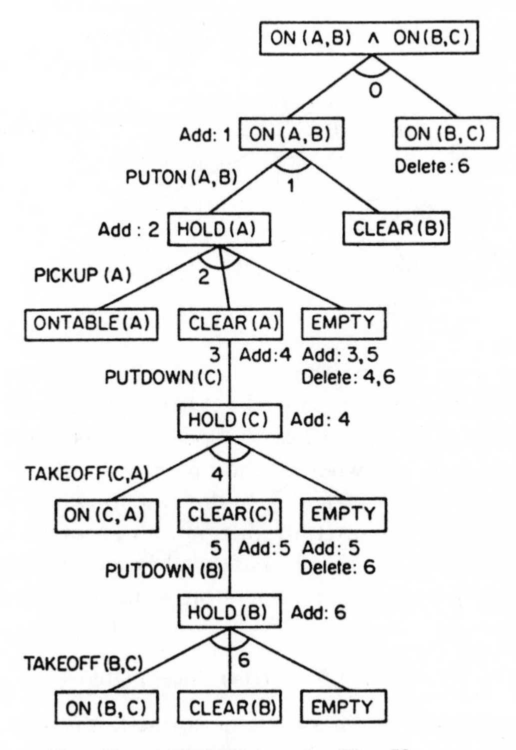

Fig. 54. AND/OR tree for Fig. 53

HOLD(A) into EMPTY. If we select, for example, PUTDOWN(A), we obtain the solution of Fig. 52. This method is more efficient than the method of finding an operator sequence by searching, as described in Section 5.4.1.

Some problems may not be solved by simple insertion of a single operator. In such cases we must construct a complete plan without contradictions by looking at goals having a final deletion operator, and successively adding operators which have this goal in their add list.

Let us consider, for instance, the solution of the problem shown in Fig. 53. A plan compiled by breaking down the goal and simply inserting operators is shown in Fig. 54. It can be seen that ON(B, C) was destroyed in order to achieve ON(A, B). We therefore add the operator PUTON(B, C), which has ON(B, C) in its add list. As shown in Fig. 55, its preconditions, HOLD(B) and CLEAR(C), become new goals. So we have to look for the addition and deletion operators of these. HOLD(B) does not satisfy the initial conditions and no addition operators appear after its deletion operator. So we cannot immediately combine this with Fig. 54. We therefore add an operator that contains HOLD(B) in its add list. If we select PICKUP(B) as such an operator, the addition and deletion operators of its preconditions are shown in Fig. 55.

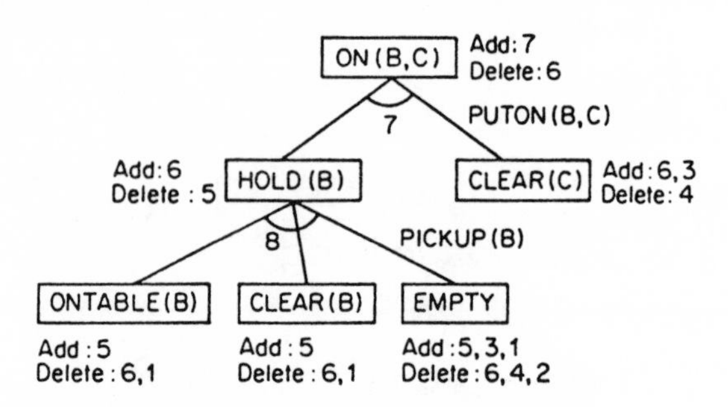

Fig. 55. Searching added operators

We can see from Fig. 55 that after application of operator 3 all of the preconditions of HOLD(B) and CLEAR(C) are satisfied. We may now, therefore, find the addition and deletion operators for each of the goals of Fig. 54 by recalculation to check whether the operator sequence of Fig. 55 can be inserted between operators 2 and 3. In this case the sequence can be inserted, and a plan in which there is no contradiction with any of the operators (operator sequence 6, 5, 4, 3, 8, 7, 2, 1) can be put together. In general, this would require further revision, and unnecessary operators would be deleted.

5.4.5 Hierarchical plans

When a complicated task has to be performed, it is often useful to devise a *hierarchical plan*, by first compiling a plan to deal with what appear to be the most important parts, then filling in the details GPS (described in Section 5.2) was a strategy involving noting the differences between a specific state and a goal state and reducing the most important of these differences. We described a strategy of selecting operators that contain in their add lists the clause representing the goal state. This strategy can also be given a hierarchical structure reflecting the importance of the various goals. We shall now explain a method of establishing a hierarchical plan by setting up a hierarchy of operator preconditions.

When, in order to achieve a given goal, we select an operator containing that goal in its add list, the preconditions (in general there will be several of these) of the operator constitute new goals. We categorize these preconditions by degree of importance and consider them in order of importance. In such a method we arrange the various goals in a hierarchy of importance, and use this to construct a rough plan for attaining the most important goals. This rough plan is then gradually refined to include the less important goals.

First of all, for comparison with GPS, consider the 'monkey and banana' problem discussed in Section 5.2. We shall use the same operators. We categorize the preconditions of the operators according to degree of importance and indicate this by a number in brackets in front of the precondition:

MOVEBOX(v)
  (1) AT(monkey, x)
  (3) AT(box, x)
CLIMB
  (1) AT(monkey, x)
  (3) AT(box, x)
GRASP
  (3) AT(box, c)
  (2) ON(monkey, box)

If we now construct a plan by considering the operators at the first level of importance (level 3), we obtain the operator sequence:

MOVEBOX(c), GRASP

If we now obtain a more detailed plan by adding the preconditions at level 2, we have:

MOVEBOX(c), CLIMB, GRASP

Lastly, the final solution is obtained by considering the preconditions at level 1:

GOTO(b), MOVEBOX(c), CLIMB, GRASP

In this example, the solution is obtained in about the same time as by using GPS. However, a precise hierarchical effect is achieved by categorizing the preconditions of each operator according to level of importance. On the other hand, in GPS we could have categorized the differences found in the general

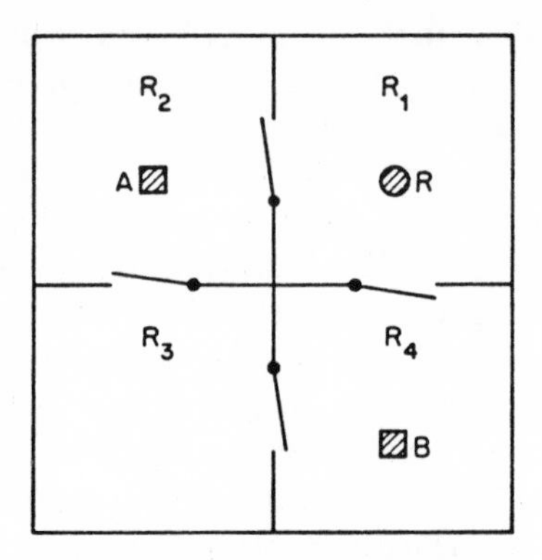

Fig. 56. Example

descriptions into hierarchical levels. Actually, formalizing the problem and selecting the operators are easier if hierarchical categorization is applied to operator preconditions rather than to differences.

The state space of a problem arrived at by considering only the preconditions of the upper levels is called an *abstract* space. The abstract space is simpler than the state space of the original problem, so much less searching is required to obtain a solution. Also it takes less time to find a lower-level solution based on a higher-level one than it does to construct a plan working at the lower level from the start. The effectiveness of using such a hierarchy of plans can be seen in the following example.

The problem is shown in Fig. 56. It is for a robot R located in a room R_1 to put boxes A and B located in rooms R_2 and R_4 next to each other in room R_3. There are doors $D(R_i, R_j)$ between rooms R_i and R_j. The problem may be represented as follows:

(i) *Initial state*
INROOM(R, R_1), INROOM(A, R_2), INROOM(B, R_4),
~OPEN(D(R_1, R_2)), ~OPEN(D(R_2, R_3)), ~ OPEN(D(R_3, R_4)),
~OPEN(D(R_4, R_1)).

(ii) Goal state
INROOM(A, R_3),
NEXTTO(A, B); this expresses that A is next to B.

(iii) *Operators and levels of preconditions*
GOTHRU(r_1, r_2); R goes from room r_1 to room r_2.
preconditions: (3) INROOM(R, r_1)
(2) OPEN(D(r_1, r_2))
delete list: INROOM(R, r_1), NEXTTO(R, $); $ indicates anything at all.
add list: INROOM(R, r_2).

GOTOB(u); R goes to the box u.
  precondition: $(\exists r)$ {(3) INROOM(R, r) ∧ (3) INROOM(u, r)}
  delete list: NEXTTO(R, $)
  add list: NEXTTO(R, u).
GOTOD(D(r_1, r_2)); R goes to D(r_1, r_2).
  precondition: (3) INROOM(R, r_1) ∨ (3) INROOM(R, r_2)
  delete list: NEXTTO(R, $)
  add list: NEXTTO(R, D(r_1, r_2)).
OPEN(D(r_1, r_2)); open the door.
  precondition: (1) NEXTTO(R, D(r_1, r_2))
  delete list: ~OPEN(D(r_1, r_2))
  add list: OPEN(D(r_1, r_2)).
PUSHTHRU(u, r_1, r_2); the robot pushes the box u from room r_1 to room r_2.
  preconditions: (3) INROOM(R, r_1)
    (4) INROOM(u, r_1)
    (1) NEXTTO(R, u)
    (2) OPEN(D(r_1, r_2))
  delete list: INROOM(R, r_1), INROOM(u, r_1), NEXTTO(u, $)
  add list: INROOM(R, r_2), INROOM(u, r_2), NEXTTO(R, u).
PUSHB(u_1, u_2); push box u_1 next to box u_2.
  precondition: $(\exists r)$ {(4) INROOM(u_1, r) ∧ (4) INROOM(u_2, r) ∧ (3) INROOM(R, r)}
    (1) NEXTTO(R, u_1)
  delete list: none
  add list: NEXTTO(u_1, u_2).

(iv) *Axioms*
OPEN(D(r_1, r_2)) ≡ OPEN(D(r_2, r_1)).
NEXTTO(u_1, u_2) ≡ NEXTTO(u_2, u_1).

The results obtained by first solving the high-level abstract space of this problem and then dropping down to progressively lower levels are as follows:

Level 4 plan

The process of plan compilation taking into account only the level 4 preconditions is shown in Fig. 57. Auxiliary nodes are inserted in the figure because there are two operators that satisfy NEXTTO(A, B). We can see from the add operators and deletion operators of NEXTTO(A, B) that the operator 1 must always be applied before operator 2 or 3. Thus at the time-point when operator 2 or 3 is applied, we have INROOM(A, R_3). Because of this, we can see that the variables r_1 and r_2 of the precondition of operators 2 and 3 must

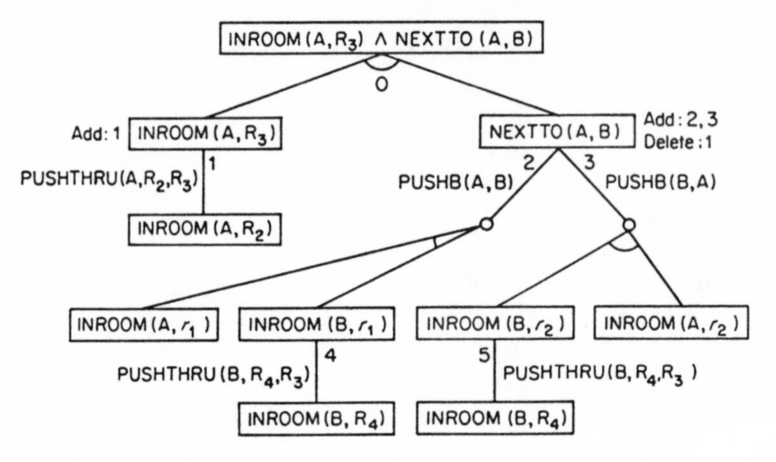

Fig. 57. AND/OR tree in abstract space of level 4

both be R_3. Operators 4 and 5 are therefore determined as shown in the figure. We have the following two solutions at level 4:

$$\text{PUSHTHRU}(A, R_2, R_3) \rightarrow \text{PUSHTHRU}(B, R_4, R_3) \begin{array}{l} \nearrow \text{PUSHB}(A, B) \\ \searrow \text{PUSHB}(B, A) \end{array}$$

Both these solutions are correct, but it will become clear at level 1 that the latter uses fewer operators. Otherwise there is scarcely any difference between them. So here we shall expand the latter.

Level 3 plan

If we refine the level 4 plan by adding the preconditions of level 4, we obtain Fig. 58. In this figure the operator sequence 1, 3, 2 at level 4 is already given, so the searching time is decreased if this is used as a basis to construct the plan. The operator sequence of the solution is 4, 1, 5, 3, 2. That is:

$$\begin{aligned} \text{GOTHRU}(r_1, R_2) &\rightarrow \text{PUSHTHRU}(A, R_2, R_3) \\ &\rightarrow \text{GOTHRU}(R_3, R_4) \rightarrow \text{PUSHTHRU}(B, R_4, R_3) \\ &\rightarrow \text{PUSHB}(B, A) \end{aligned}$$

where r_2 in Fig. 58 is R_3, and r_1 is any room (fixed at R_1 in the level 1 plan).

Level 2 plan

When we compile a more detailed plan in the same way, the following plan is obtained (the reader should verify this):

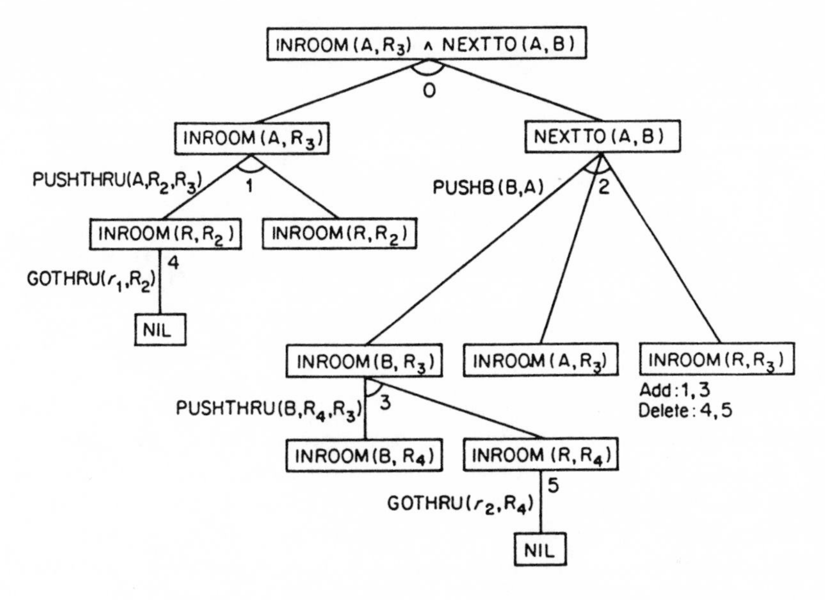

Fig. 58. AND/OR tree in abstract space of level 3

$$\text{OPEND}(\text{D}(r_1, \text{R}_2)) \rightarrow \text{GOTHRU}(r_1, \text{R}_2) \rightarrow \text{OPEND}(\text{D}(\text{R}_2, \text{R}_3))$$
$$\rightarrow \text{PUSHTHRU}(\text{A}, \text{R}_2, \text{R}_3) \rightarrow \text{OPEND}(\text{D}(\text{R}_3, \text{R}_4))$$
$$\rightarrow \text{GOTHRU}(\text{R}_3, \text{R}_4) \rightarrow \text{PUSHTHRU}(\text{B}, \text{R}_4, \text{R}_3) \rightarrow \text{PUSHB}(\text{B}, \text{A})$$

Level 1 plan

We leave the process of obtaining the solution to the reader. The result is as follows:

$$\text{GOTOD}(\text{D}(\text{R}_1, \text{R}_2)) \rightarrow \text{OPEND}(\text{D}(\text{R}_1, \text{R}_2)) \rightarrow \text{GOTHRU}(\text{R}_1, \text{R}_2)$$
$$\rightarrow \text{GOTOD}(\text{D}(\text{R}_2, \text{R}_3)) \rightarrow \text{OPEND}(\text{D}(\text{R}_2, \text{R}_3)) \rightarrow \text{GOTOB}(\text{A})$$
$$\rightarrow \text{PUSHTHRU}(\text{A}, \text{R}_2, \text{R}_3) \rightarrow \text{GOTOD}(\text{D}(\text{R}_3, \text{R}_4))$$
$$\rightarrow \text{OPEND}(\text{D}(\text{R}_3, \text{R}_4)) \rightarrow \text{GOTHRU}(\text{R}_3, \text{R}_4) \rightarrow \text{GOTOB}(\text{B})$$
$$\rightarrow \text{PUSHTHRU}(\text{B}, \text{R}_4, \text{R}_3) \rightarrow \text{PUSHB}(\text{B}, \text{A})$$

This involves a fairly long sequence of operators, which it would be inefficient to compile at a single level. The key to efficient compilation of a hierarchical plan lies in proper selection of the levels of the preconditions.

Chapter 6

Languages for Artificial Intelligence

In earlier chapters we have considered methods of solving clearly defined problems by searching. These methods may provide part of the solution of larger artificial intelligence problems such as theorem-proving, natural-language processing, and image-processing. Artificial intelligence problem-solving systems are, of course, expressed in computer programming languages, so this gives rise to a requirement for languages that are as convenient as possible for this purpose.

Often when computers have been applied to new fields, new programming languages specially suited to those fields have been developed. The problem-oriented languages so far developed include languages for use in numerical computation, business computation, and simulation. Large systems can be constructed using problem-oriented programming languages, since they enhance programming efficiency.

From the earliest stages of research into artificial intelligence, languages specially suited to this purpose have been developed; and for a long time the language which has been most used is LISP. LISP is a general-purpose language, but it is well suited to artificial intelligence since it enables complicated computing procedures, especially symbol manipulation procedures, to be expressed in a simple way. Other programming languages based on LISP but with improved facilities were developed in the 1970s, and include PLANNER, CONNIVER and POP-2. Another, called INTER-LISP, was developed at about this time, and is still used. It is based on LISP, with various facilities added and improved.

The programming languages QA1, QA2, and QA3 were devised at the end of the 1960s for theorem-proving using predicate calculus. Under the influence of PLANNER, they have now been modified to QA4 to incorporate procedural representation. If too many facilities are included, the system becomes unmanageable, slowing down the processing speed and making debugging difficult. The language PROLOG has therefore been proposed, which uses predicate calculus to make inferences. It is now used in various research establishments, and is being evaluated.

In this chapter we shall give a detailed account of LISP and an example of a problem-solving program using it. We shall also outline the typical artificial intelligence languages PLANNER, CONNIVER, and PROLOG. The languages of knowledge representation will be discussed in Chapter 7.

6.1 LISP

LISP is the language which has been most used in artificial intelligence since it was developed in the late 1950s by J. McCarthy of MIT. LISP is more suitable for carrying out symbolic processing or flexible control than were the previously used programming languages, which were intended for numerical computation. It is a general-purpose language, but, with progress of research into artificial intelligence, it has been improved and now affords a faster processing speed, various facilities for program debugging and conversational processing. Apart from artificial intelligence, it is now used for formula manipulation and editing, etc. Many of the languages for artificial intelligence described later in this chapter are written using LISP.

6.1.1 Structure of programs and data

In LISP, both programs and data have the same structure. The basic elements of this structure are sequences of characters called *atoms*. An atom may consist of any sequence of characters apart from a blank. Integers and floating-point numbers are also atoms. Examples are given below:

```
A            DEPTH-FIRST-SEARCH
JOHN         123
?X           3.141592
LONGATOM3
```

An expression consisting of any number of atoms (including 0) surrounded by brackets is called a *list*. The atoms that make up the list are called *elements* of the list. In general, the elements can include not only atoms but also lists. Examples of lists are given below:

```
(A B C)   ()
(A)       (SYMBOL (INTEGER REAL))
```

The general term for atoms and lists is *S-expressions* (symbolic expressions). LISP programs and data are both represented by S-expressions.

6.1.2 Basic functions

Atoms other than numbers can be given a value. For example, if we want to give the atom TWO the value 2, we do this as follows:

```
(SETQ TWO 2)
```

SETQ is a basic LISP function. It has two arguments, and gives the value of the second argument as the value of the first argument. The second argument is the number 2, so its value is 2. So the value of TWO becomes 2. Finding the value of an atom is called *evaluating* the atom. In the LISP system, when an atom is given it is evaluated. So if we input TWO to the system, we obtain 2 as the response. This is written:

```
TWO
*2
```

The asterisk indicates a response from the system.

In general, S-expressions have a value. For example, the list (PLUS 2 3) that brings into play the function PLUS (which performs addition) has the value 5:

```
(PLUS 2 3)
*5
```

If we do not want an S-expression to be evaluated, we attach the quotation mark:

```
'TWO
*TWO
'(PLUS 2 3)
*(PLUS 2 3)
```

With the above preparations, we can introduce functions that operate on lists. For example, CAR returns (as a value) the head element of a list, and CDR returns the list with the head element deleted:

```
(CAR '(FIRST SECOND THIRD))
*FIRST
(CAR '((A B)C))
*(A B)
(CDR '(FIRST SECOND THIRD))
*(SECOND THIRD)
(CDR '(ONE))
*( )
```

( ) indicates the empty list. CONS takes two arguments and inserts the first argument at the head of the list of the second argument:

```
(CONS 'A '(B C))
*(A B C)
(CONS (CAR '(A B C))(CDR '(A B C)))
*(A B C)
```

A function may appear as the argument of a function. The list is evaluated taking the items in the innermost brackets first. Atoms without quotation marks are evaluated in the same way:

```
(SETQ L '(A B C))
*(A B C)
(CAR L)
*A
(CONS L L)
*((A B C)A B C)
```

It should be noted that although the S-expression in the first line above has the value (A B C), the object of the function SETQ is not to get the value, but to give this value to L. Such a change produced when a function is evaluated is called a *side effect.*

LIST takes any number of arguments and produces a list with them as elements:

```
(LIST (LIST 'A 'B)(LIST 'C 'D))
*((A B)(C D))
```

APPEND joins up lists:

```
(APPEND '(A B)'(C D))
*(A B C D)
```

ATOMCAR and ATOMCDR are used to handle sequences of letters:

```
(ATOMCDR 'YAMADA)
*Y
(ATOMCDR 'YAMADA)
*AMADA
```

SET is a function which has two arguments. It evaluates the first argument and gives the value of the second argument as the resulting value:

```
(SET (CAR '(X Y))'(A B))
*(A B)
```

Thus what happens when the first line is evaluated is that X is found as the value of (CAR '(X Y)), and this X is in turn given the value (A B):

```
X
*(A B)
(SET (ATOMCDR 'SPOT)5)
*5
POT
*5
```

SETQ can be regarded as a combination of SET and ‘. For example, the value and side-effects of (SET ‘Z ‘(C D)) and (SETQ Z ‘(C D)) are the same.

LISP has the numerical computation functions PLUS, DIFFERENCE, TIMES, QUOTIENT, SQRT, ABS, and MAX, etc.

6.1.3 Predicate functions

A predicate function is a boolean function that returns the values *true* and *false*. In LISP, true and false are expressed by special atoms T and NIL. The values of T and NIL are respectively T and NIL. The basic predicate functions will now be given.

ATOM returns the value T if its argument is an atom, and otherwise returns the value NIL:

```
(ATOM ‘YAMADA)
*T
(ATOM ‘(THIS IS A LIST))
*NIL
```

Using the L (= ‘(A B C)) defined in the explanation of CONS, we have:

```
(ATOM L)
*NIL
(ATOM ‘L)
*T
```

ZEROP returns T if its argument is 0, otherwise it returns NIL:

```
(ZEROP 2)
*NIL
```

EQUAL returns T if its first and second arguments are equal, but if they are not equal returns NIL:

```
(EQUAL L L)
*T
(EQUAL L ‘(A B C))
*T
(EQUAL L ‘L)
*NIL
```

NULL checks to see whether the argument is or is not the empty list:

```
(NULL ‘( ))
*T
(NULL L)
*NIL
```

```
(NULL (CDR(CDR '(A B))))
*T
```

MEMBER checks to see whether the first argument is an element of the second argument:

```
(MEMBER 'A L)
*T
(MEMBER 'A (CDR L))
*NIL
```

The functions that perform boolean computation are NOT, AND, and OR:

```
(NOT (ATOM 'A))
*NIL
(NOT (ATOM L)
*T
(AND (NOT (NULL L))(ATOM(CAR L)))
*T
(AND T (ATOM L))
*NIL
(OR (NULL L)(ATOM L))
*NIL
(OR (NULL L)(ATOM (CAR L)))
*T
```

OR and AND can take any number of arguments:

```
(AND T T NIL)
*NIL
```

6.1.4 Evaluation depending on conditions

When we want to change the subject of evaluation depending on some condition, we use the special function COND. In general COND has the following form:

```
(COND(⟨test 1⟩⟨subject of evaluation 1⟩)
      (⟨test 2⟩⟨subject of evaluation 2⟩)
                        .
                        .
                        .
(⟨test n⟩⟨subject of evaluation n⟩))
```

First the LISP system evaluates ⟨test 1⟩. If it is T, ⟨subject of evaluation 1⟩ is evaluated, and the procedure terminates. But if ⟨test 1⟩ is NIL, the system goes to the next line. Finally, if ⟨test *n*⟩ is NIL, NIL is returned. If when a ⟨test *i*⟩ is evaluated the result is T, but there is no ⟨subject of evaluation *i*⟩, T is returned and the procedure terminates. COND is often used in defining functions, but is scarcely ever used on its own.

6.1.5 Definition of functions

The user can freely define functions apart from the basic functions provided by LISP described above. Functions are defined using the function DEFUN. DEFUN is used as follows:

```
(DEFUN⟨name of function⟩⟨list of parameters⟩
⟨definition of function⟩)
```

Using this function, we can define the function SECOND that finds the value of the second element in a list:

```
(DEFUN SECOND (X)(CAR(CDR X)))
*SECOND
```

Once the function SECOND has been defined, this definition is valid until the function is redefined:

```
(SECOND L)
*B
```

The procedure that LISP uses to evaluate the above S-expression is as follows. When (SECOND L) is inputted, the system searches for the definition of SECOND. There is one formal parameter in the definition, namely X. The system therefore stores the previous value which X had before this function was entered in the buffer region, and now connects X with L. This is called binding X to L. With this binding, the definition of the function is evaluated, taking X as L. That is, (CAR(CDR L)) is computed. The result is B. Next the binding of X is removed, and the value of X that was stored in the buffer region is restored. Finally the value B is returned.

The function UPDATE-OPEN that is used in searching a graph to add a newly found node N to the OPEN list if it is not already there is defined as follows:

```
(DEFUN UPDATE-OPEN(NEW-NODE)
  (COND((MEMBER NEW-NODE OPEN) OPEN)
         (T (CONS NEW-NODE OPEN))))
```

If now OPEN is (A B C D), we have:

```
(UPDATE-OPEN 'E)
*(E A B C D)
```

```
(UPDATE-OPEN 'C)
*(E A B C D)
```

6.1.6 Recursive call and iteration

It is convenient for a function to be able to call itself. Let us define the function MEMBER referred to above as an example of a function for which recursive call is performed:

```
(DEFUN MEMBER (A L)
  (COND ((NULL L) NIL)
        ((EQUAL A (CAR L)) T)
        (T (MEMBER A (CDR L)))))
```

If we now input (MEMBER 'BLUE '(RED GREEN BLUE YELLOW)). A is bound to BLUE and L is bound to (RED GREEN BLUE YELLOW). Next the computer executes the definition portion. L is not the empty list and A is not the same as (CAR L), so the last line (MEMBER A (CDR L)) is evaluated. In other words, the given problem is reduced to (MEMBER 'BLUE '(GREEN BLUE YELLOW)). When this is evaluated, (MEMBER 'BLUE '(BLUE YELLOW)) is again called. The system then finds that (EQUAL 'BLUE (CAR '(BLUE YELLOW))) holds, so the result is T. So the original function application also returns T. The above process is shown in Fig. 59.

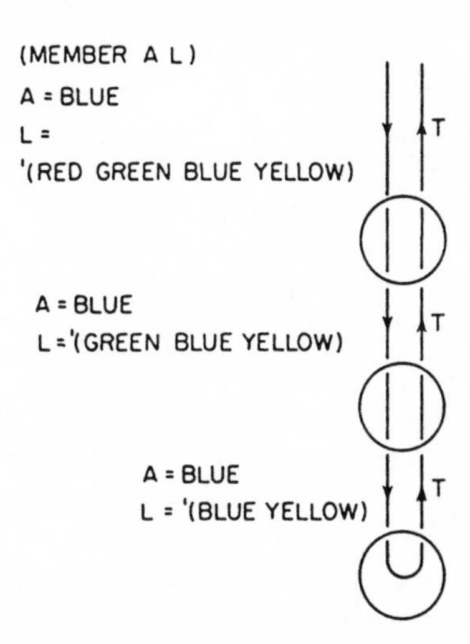

Fig. 59. Process of execution of MEMBER. Circles indicate the function MEMBER, and the values of the arguments and the evaluation values of the functions are next to the arrows

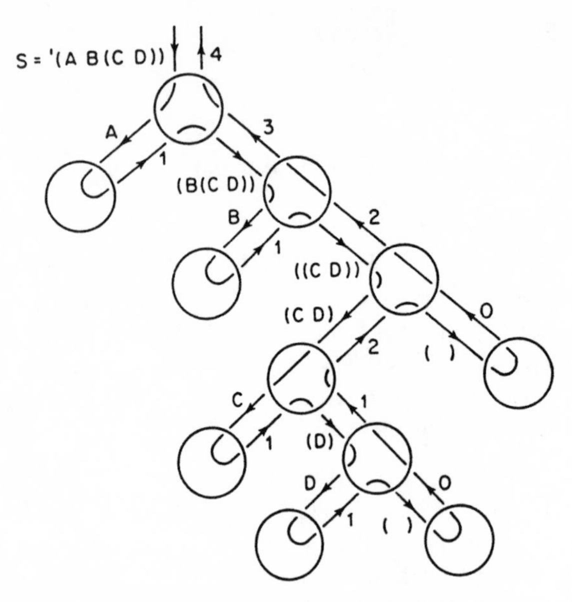

Fig. 60. Process of execution of COUNTATOM

If we input (MEMBER 'PINK '(RED GREEN BLUE YELLOW)), when the above evaluation is carried out, L ends up as being the empty list, so NIL is returned.

As a further example, let us define a function that finds the number of atoms in an S-expression:

```
(DEFUN COUNTATOM (S)
  (COND ((NULL S) 0)
        ((ATOM S) 1)
        (T (PLUS (COUNTATOM (CAR S))
                 (COUNTATOM (CDR S))))))
```

An example of this function is:

```
(COUNTATOM '(A B (C D)))
*4
```

The process of carrying out this evaluation is shown in Fig. 60. The argument is simplified every time the function is recursively called, until an argument that can be directly evaluated is found.

LISP can also use iteration. For this purpose a procedure may be defined using the function PROG. Within PROG we can write procedures that include iteration or GOTO or successive execution of several procedures such as FORTRAN programs. We can use PROG to define the function MEMBER referred to above:

```
(DEFUN MEMBER (A L)
  (PROG (LREMAIN)
    (SETQ LREMAIN L)
    LOOP
    (COND ((NULL LREMAIN)(RETURN NIL))
          ((EQUAL A (CAR LREMAIN))(RETURN T)))
  (SETQ LREMAIN (CDR LREMAIN))
  (GO LOOP)))
```

Immediately after PROG, we define a *local* variable (in this case LREMAIN) that is only used within PROG. LOOP is an atomindicating location, so that control can be transferred there by (GO LOOP). If the control reaches (RETURN⟨S-expression⟩), PROG terminates, and returns as its value the value of the immediately following S-expression. In this example the value will be either T or NIL.

Iteration has the advantage that it usually requires less computer memory since the function need not be called many times. However, in cases such as that of the second example COUNTATOM, where two or more recursive calls are carried out at once, the procedure cannot be expressed by simple iteration.

Recursive calling can occur even if the function does not directly call itself. For example, we still have recursion if there are functions F1, F2, and F3 and F1 calls F2, which calls F3, which in turn calls F1. A specific example of this is shown in Section 6.2. It is difficult to write such processing without using recursive call.

6.1.7 Properties

A symbolic atom (i.e. an atom other than a number or T and NIL) may have a property. For example, a block can have the property of color. The property values are the names of the various colours. For example, if we make the property COLOR of BOX1 be RED:

```
(PUTPROP 'BOX1 'RED 'COLOR)
*COLOR
```

When we want to know the value of the property COLOR of BOX1:

```
(GET 'BOX1 'COLOR)
*RED
```

Now suppose that there are several different types of block on the table. Write the list of blocks as LBLOCK and assume that values are given for the property COLOR of each block. The function GETBLUE that finds the blue block is defined as follows:

```
(DEFUN GETBLUE(L)
```

```
    (COND ((NULL L))
          ((EQUAL (GET (CAR L) 'COLOR) 'BLUE)(CAR L))
          (T (GETBLUE (CDR L)))))
```

Now suppose there is the following input in addition to this definition:

```
(SETQ LBLOCK '(BRICK PYRAMID BOX))
(PUTPROP 'BRICK 'GREEN 'COLOR)
(PUTPROP 'PYRAMID 'BLUE 'COLOR)
(PUTPROP 'BOX 'RED 'COLOR)
```

If we now execute GETBLUE the result is as follows:

```
(GETBLUE LBLOCK)
*PYRAMID
```

One atom may have several properties. For instance, we can arrange for BOX1 to have other properties apart from color as follows:

```
(PUTPROP 'BOX1 '6 'HEIGHT)
(PUTPROP 'BOX1 'SQUARE 'SHAPE)
```

Although we have not mentioned it up to this point, for simplicity, there are in fact several varieties of LISP, differing slightly in types of functions, the ways in which they act, and the methods of writing programs. The description in this chapter is of MACLISP, which was developed at MIT and is the dialect of LISP which is most widely used.

6.1.8 Matching

Now that we have described the basic facilities of LISP, let us write a moderately complex program. We expand EQUAL, which is a predicate function that checks to see whether two S-expressions are the same, and define the function MATCH that performs a matching allowing a little more latitude. Assume that a pattern P and S- expression S are to be matched. We want to arrange that the elements of P may include the atom ? and that ? is matched with any S-expression. For example:

```
(MATCH ((COLOR ? RED) '(COLOR BOX RED))
*T
```

To match S-expressions, we may check to see whether the elements of the two lists match. First MATCH checks to see whether the head elements match. If they do, it checks matching of the rest of the lists by recursive calling:

```
(DEFUN MATCH (P S)
  (COND ((AND (NULL P)(NULL S)) T)
        ((OR (NULL P)(NULL S)) NIL)
```

```
            ((OR (EQUAL (CAR P) '?)
                  (EQUAL (CAR P)(CAR S)))
            (MATCH(CDR P)(CDR S)))))
```

By using this MATCH function as in the preceding example, we can check to see whether there are any red objects. But if there are some red objects, we cannot tell which they are. We therefore introduce atoms beginning with ? which match any S-expression, and expand MATCH so that the matched S-expressions can be read out afterwards. The new MATCH operates as follows:

```
(MATCH '(COLOR ?X RED) '(COLOR BOX RED))
*T
X
*BOX
```

?X matched BOX, so BOX was returned as the value of X. To achieve this, we can insert the following procedure into the COND of MATCH:

```
((AND (EQUAL (ATOMCAR (CAR P)) '?)
      (MATCH (CDR P)(CDR S)))
(SET (ATOMCDR (CAR P))(CAR S))
T)
```

That is, if the head element of P starts with ?, this procedure checks to see whether the rest matches. If it does match, it sets the head element of the S-expression as the value of the atom expressed by the character sequence following the ?, and returns T as the value of the function. (Note that it does not matter even if the subject of evaluation in COND is several S-expressions.)

With these preparations, it is simple to define a new MATCH:

```
(DEFUN MATCH (P S)
  (COND ((AND (NULL P)(NULL S)) T)
        ((OR (NULL P)(NULL S )) NIL)
        ((OR (EQUAL (CAR P) '?)
              (EQUAL (CAR P)(CAR S)))
        (MATCH (CDR P)(CDR S)))
        ((AND (EQUAL (ATOMCAR (CAR P)) '?)
               (MATCH (CDR P)(CDR S)))
        (SET (ATOMCDR (CAR P))(CAR S))
        T)))
```

It is also possible (though it will not be described here) to further expand this MATCH to give it the facility to match a pattern element to a sequence of an arbitrary number of atoms.

Now consider the problem of using match to look up facts from a database, or to update the database. The database is a list of facts, so let us represent it by the variable LDATA. To start with, the initial value is given by:

```
(SETQ LDATA '((ON A C)(ON B D)(ONTABLE C)
                                  (ONTABLE D)))
```

A function to look up facts from the database may be defined as follows:

```
(DEFUN FIND (FACT)
  (PROG (LTEMP)
     (SETQ LTEMP LDATA)
     LOOP
     (COND ((NULL LTEMP)(RETURN NIL))
           ((MATCH FACT (CAR LTEMP))(RETURN T)))
     (SETQ LTEMP (CDR LTEMP))
     (GO LOOP)
```

For example, with LDATA referred to above:

```
(FIND '(ON ?X C))
*T
X
*A
(FIND '(ON ?X A))
*NIL
```

A function to remove a fact from a database can be defined as follows using an auxiliary function DELETE1:

```
(DEFUN DELETE (FACT)
  (SETQ LDATA (DELETE1 FACT LDATA)))
(DEFUN DELETE1 (FACT L)
  (COND ((NULL L) NIL)
        ((MATCH FACT (CAR L))(CDR L))
        (T (CONS (CAR L)(DELETE1 FACT (CDR L))))))
```

We give an example of DELETE below:

```
(DELETE '(ON B D))
*((ON A C)(ONTABLE C)(ONTABLE D))
```

A function to add a fact to a database may be defined as follows:

```
(DEFUN ADD (FACT)
  (SETQ LDATA (CONS FACT LDATA)))
```

For example:

```
(ADD '(ONTABLE B))
*((ONTABLE B)(ON A C)(ONTABLE C)(ONTABLE D))
```

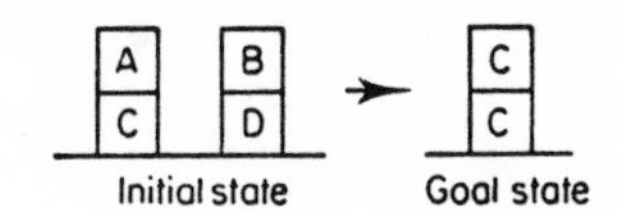

Fig. 61. Blocks problem

6.2 PROBLEM-SOLVING BY MEANS OF PROCEDURES

We shall now show how LISP is used to write programs to solve specific problems. A typical simple problem is the blocks problem dealt with in Section 5.4. The problem is shown in Fig. 61. Let us assume that the state of the problem can be expressed by the database LDATA referred to previously. The initial state is represented by the following list:

```
((ON A C)(ON B D)(ONTABLE C)(ONTABLE D)(EMPTY))
```

The problem is to make the LDATA contain (ON C D) by moving the blocks. Procedures corresponding to the operators have already been given. For simplicity we shall use three procedures. Their effects and definitions of their functions are given below.

(PUTON X Y)

The block X is put on top of the block Y. If X is already on top of Y, this does nothing. If there is anything on top of Y it takes it off, puts it on the table, and puts X on top of Y. The function (PUTAT B A) is used to put a block B at a place A (which could be either a block or the table) on which nothing is standing. This is expressed as follows by a program:

```
(DEFUN PUTON (X Y)
  (PROG (?Z Z)
    (COND ((FIND (LIST 'ON X Y))(RETURN)))
    LOOP
    (COND ((FIND (LIST 'ON ?Z Y))
              (PUTAT Z 'TABLE)(GO LOOP)))
    (PUTAT X Y)))
```

Obviously if there are several blocks on top of Y they will be taken off one after the other.

(PUTAT X Y)

This is a function that puts block X on top of Y. It is understood there is nothing on top of Y. If X is already grasped, it is placed on Y. If it is not already grasped, (PICKUP X) is used to grasp X. When the operation is completed, LDATA is updated:

```
(DEFUN PUTAT (X Y)
  (PROG (?Z Z)
    (COND ((NOT (FIND (LIST 'HOLD X)))
      (COND ((FIND (LIST 'HOLD ?Z))
               (PUTAT Z 'TABLE)
               (DELETE (LIST 'HOLD Z))
               (ADD (LIST 'ONTABLE Z))))
      (PICKUP X)))
    (DELETE (LIST 'HOLD X))
    (COND ((EQUAL Y 'TABLE)(ADD (LIST 'ONTABLE X)))
          (T (ADD (LIST 'ON X Y))))
    (RETURN)))
```

(PICKUP X)

This picks up X. If there is anything on X it is put on the table and then is picked up. Finally the state is updated:

```
(DEFUN PICKUP (X)
        (PROG (?Y Y)
        LOOP
        (COND ((FIND (LIST 'ON ?Y X))
                (PUTAT Y 'TABLE)(GO LOOP)))
        (COND ((FIND (LIST 'ON X ?Y)(DELETE (LIST 'ON X Y)))
                (T (DELETE (LIST 'ONTABLE X)))))
        (ADD (LIST 'HOLD X))
        (RETURN)))
```

If there are several blocks on X, obviously they are taken off one after another. As shown in Fig. 62, if Y is on top of X and Z is on top of Y, they are taken off in order from the top. When (PICKUP A) is applied in the state shown at the top left of the drawing, the state is changed by successive calling of the functions. It can be seen that PICKUP and PUTAT successively call each other. Thus the function calls itself through calling another function. This is also an example of recursion.

Returning to the initial problem shown in Fig. 61, we try to reach (ON C D) by running (PUTON C D). The flow of control is shown in Fig. 63. Any conditions which are necessary before the function can be called are shown at the side of the arrows. The state change achieved by the function is also shown. It should be noted that the cancellation of (ON X Y) is expressed by (CLEAR Y) and the cancellation of (HOLD X) is expressed by (NOT (HOLD X)).

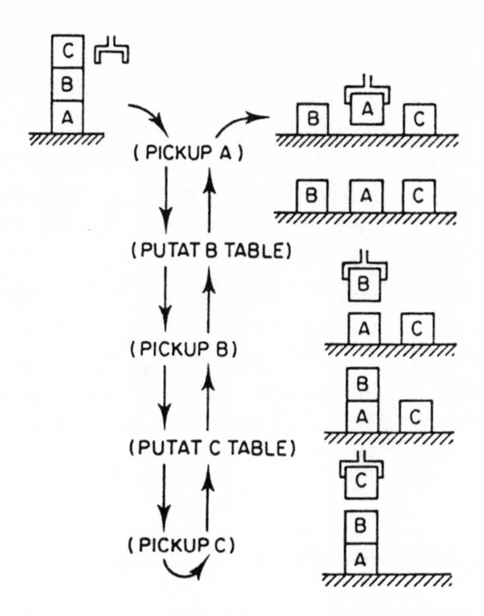

Fig. 62. Example of the operation of PICKUP

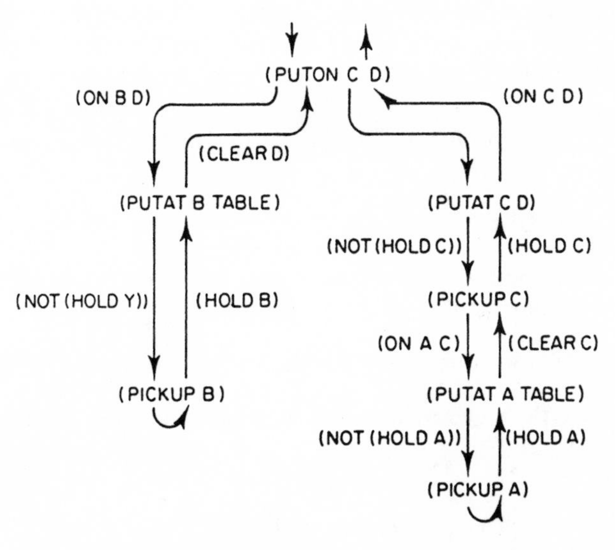

Fig. 63. Procedure to reach (ON C D)

Thus fairly complicated processing can be achieved by compiling procedures for problem solution. Though this problem does not require searching, problem-solving involving searching can also be done using LISP.

6.3 *PLANNER*

PLANNER is a high-level computer language devised for problem-solving or theorem-proving. With previous computer languages it was necessary to indicate directly what procedure was to be used for solving a problem. In contrast, if we input beforehand the necessary data and procedures, PLANNER will automatically search the data and procedures that can be used to solve a given problem, and provide an answer.

The basic concept of PLANNER was published by C. Hewitt, a student at MIT, in 1968. Many later improvements have been made. One such development is MICRO-PLANNER, which incorporates some of the facilities of PLANNER and which became available in 1970. We shall describe here the outlines of MICRO-PLANNER, since it in fact incorporates almost all of the essential facilities of PLANNER.

6.3.1 Basic operation of PLANNER

To facilitate understanding of PLANNER, we shall describe its basic operation with reference to examples. PLANNER has a fact database and a procedure database. The fact database stores facts corresponding to the state of the problem. The procedure database stores procedures corresponding to operators. Suppose now there are on the table rectangular parallelepipeds BLOCK1 and BLOCK2, and square pyramids PYRAMID1 and PYRAMID2. We express the fact that BLOCK1 is a rectangular parallelepiped (block) by (BLOCK BLOCK1). We insert this fact in the fact database as follows:

```
(THASSERT (BLOCK BLOCK1))
```

The TH at the beginning of a function name shows that it is a PLANNER function. To input the fact that BLOCK1 is red:

```
(THASSERT (COLOR BLOCK1 RED))
```

We also input into the fact database that BLOCK2 and PYRAMID1 are blue, and PYRAMID1 is red.

If we want to find a block, we can input:

```
(THPROG (X)(THGOAL (BLOCK ?X)))
```

THPROG (X) corresponds to $\exists X$, and means finding an arbitrary X. PLANNER finds a pattern in the fact database that matches (BLOCK ?X). (As explained in Section 6.1, ?X matches any atom.) If the pattern that it finds first is (BLOCK BLOCK1), X is given the value BLOCK1. If next we want to find a blue block, we input the following:

```
(THPROG (THGOAL (BLOCK ?X))
        (THGOAL (COLOR $X BLUE)))
```

This means that two goals are to be satisfied at the same time. First the computer executes (THGOAL (BLOCK ?X)), makes the value of the X obtained $X, and thereby attains the second goal. In this example the value of X initially becomes BLOCK1. Next PLANNER looks for (COLOR BLOCK1 BLUE), but does not find it. So PLANNER returns to the immediately preceding decision, and repeats the search. It then finds that (BLOCK ?X) matches (BLOCK BLOCK2). Now the second goal (COLOR BLOCK2 BLUE) is attained, X is confirmed as BLOCK2, and the procedure terminates. As can be seen from the above examples, PLANNER variables must start with ? or $. Otherwise they are regarded as starting with the quotation mark ‘.

Next, if we want to define a theorem ‘the upper surface of a block is flat’, we define the following procedure:

```
(THCONSE (X) (FLAT ?X)
             (THGOAL (BLOCK ?X)))
```

That is, to reach the main goal X is FLAT, we should attain the sub-goal that X is a block. This procedure is stored in the procedure database.

If now we want to find a red building block having a flat top, we may proceed as follows:

```
(THPROG (THGOAL (FLAT ?X))
        (THGOAL (COLOR $X RED)))
```

First of all, PLANNER searches the fact database for a fact that matches (FLAT ?X). In this case there is no such fact, so it searches the procedure database for a procedure that has as its goal (FLAT ?X). When it finds the procedure given earlier, it replaces (FLAT ?X) by the sub-goal (BLOCK ?X) for attaining this goal. In the subsequent steps, the value of X becomes BLOCK1, as in the preceding example.

6.3.2 Advantages of PLANNER

As we saw from the explanation of its basic operation, PLANNER has the following three advantages:

(1) The examination of the fact database is performed by means of a pattern. Given a list of atoms containing variables, facts matching it can be obtained automatically.
(2) Procedures are called by a pattern. A procedure matching a goal is automatically called simply by specifying the goal, without needing to specify the name of the procedure.

(3) Searching is performed automatically. PLANNER selects from the database suitable facts or procedures required to attain the goal. The items selected will not necessarily be correct; but if this results in failure of the search, PLANNER will back up to the immediately preceding decision-points and then select again.

We shall show how the searching is carried out using a rather more complicated example. Assume there are some building blocks additional to the building blocks mentioned previously, and that the following facts are inputted concerning them:

```
(THASSERT (CYLINDER CYLINDER1))
(THASSERT (CYLINDER CYLINDER2))
(THASSERT (COLOR CYLINDER1 BLUE))
(THASSERT (COLOR CYLINDER2 GREEN))
```

We also assume that the following procedure is given:

```
(THCONSE (X) (FLAT ?X)
             (THGOAL (CYLINDER ?X)))
```

If now we want to find a green building block with a flat top, we can write the following procedure:

```
(THPROG (THGOAL (FLAT ?X))
        (THGOAL (COLOR $X GREEN)))
```

As in the previous example, since there is no data matching (FLAT ?X), this goal is replaced by the sub-goal (BLOCK ?X) of the procedure (used in the previous example) that has (FLAT ?X) as its goal. However, no data is found that satisfies (BLOCK $X) and (COLOR $X GREEN), and there is no procedure that has these as goal. PLANNER therefore searches for another procedure that has (FLAT ?X) as goal. When it does this, it finds the procedure given above. It then replaces (FLAT ?X) by (CYLINDER ?X). Now since (CYLINDER ?X) and (COLOR $X GREEN) are both true at the same time, the value of X becomes CYLINDER2.

Thus PLANNER will automatically perform a backward depth-first search on a fact database and procedure database.

6.3.3 Database management

In Section 6.3.1 we described a method whereby facts and procedures could be stored in a database. In addition to this, PLANNER offers other means of database management. One of these is to define a procedure using the function THANTE. THCONSE specifies which sub-goal must be attained in order to achieve a given goal. That is, it is called by the consequent. In contrast,

THANTE is called by the antecedent. For example, if we want to store in the fact database the fact that a building-block is a rectangular parallelepiped with a flat top, we can register the following procedure:

```
(THANTE (X)(BLOCK ?X)
        (THASSERT (FLAT $X)
```

The result of this is that if we enter the fact (BLOCK BLOCK1) onto the database by inputting:

```
(THASSERT (BLOCK BLOCK1))
```

then (FLAT BLOCK1) is automatically generated. This gives a gain in efficiency because we do not need to call the THCONSE procedure every time (FLAT ?X) is used if (FLAT ?X) is used several times. However, it does have the drawback of increasing the number of facts in the database if there are a large number of rectangular parallelepipeds, since the fact that the parallelepiped is FLAT will be generated in the case of each of them.

THERASING defines a procedure that cancels a fact from the database. For example, if (BLOCK X) has been eliminated from the database, and we want to delete (COLOR X Y), we may register the following procedure:

```
(THERASING (X Y)(BLOCK ?X)
           (THERASE (COLOR $X ?Y)))
```

If we want to do the same thing for all the building blocks, not just the rectangular parallelepipeds:

```
(THERASING (X Y Z)(?X ?Y)
           (THERASE (COLOR $Y ?Z)))
```

It should be noted that, if various types of facts are stored in the fact database, care should be taken that this does not have unintended results, since (?X ?Y) matches any list of two elements.

Databases can also be manipulated by a THCONSE-type procedure. For example, if the robot grasps a building block X, if immediately previously we had (ON X Y), we remove this fact, and add (HOLD X). This is achieved as follows (for simplicity, the precondition CLEAR (X) is omitted):

```
(THCONSE (X Y)(HOLD ?X)
  (THGOAL (ON ?X ?Y))
  (THERASE (ON $X $Y))
  (THASSERT (HOLD $X)))
```

In this way, we can write a processing procedure for attaining a goal during a procedure.

6.4 CONNIVER

CONNIVER is an artificial intelligence language developed at MIT in 1972 by G. J. Sussman, one of the creators of MICROPLANNER, as an extension from PLANNER.

On many points CONNIVER carries forward the characteristics of PLANNER; the factual database, procedural database and pattern-based database search have been adopted without change.‡ The principal difference is that automatic backtracking is abandoned, and control of the problem-solving is left to the user. This is because, depending on the problem, the unconditional depth-first search carried out by PLANNER may be inefficient. In addition to this change a number of functions are enhanced, and these will be discussed next.

6.4.1 Provision of flexible control

First we shall discuss a simple example in which automatic backtracking is inappropriate. Suppose that the initial state of the factual database is:

```
(AT PERSON1 HOME)
(AT MONEY HOME)
```

and the target state is:

```
(GETON PERSON TRAIN)
```

and also suppose that the following procedure is given:

```
1. (THCONSE (X)(GETON ?X TRAIN); catch a train
     (THGOAL (AT ?X STATION))
     (THGOAL (HAVE $X TICKET))
     (THERASE (AT $X STATION))
     (THASSERT (GETON $X TRAIN)))
```

If the meaning of the procedure is understood, the THERASE and THASSERT portions are self-evident, and will be omitted in the following.

```
2. (THCONSE (X)(AT ?X STATION); go to station
     (THGOAL (AT ?X BUS-STOP)))

3. (THCONSE (X)(AT ?X BUS-STOP); go to bus-stop
     (THGOAL (AT ?X HOME)))
```

‡CONNIVER function names are completely different from those of PLANNER. Procedures are given names, and for example THCONSE in PLANNER becomes (IF-NEEDED function name) in CONNIVER. Discussion of function names will be omitted here.

```
4. (THCONSE (X)(HAVE ?X TICKET); buy a ticket
      (THGOAL (AT ?X STATION))
      (THGOAL (HAVE $X MONEY)))

5. (THCONSE (X Y)(HAVE ?X MONEY)
      (THGOAL (AT MONEY ?Y))
      (THGOAL (AT ?X $Y)))
```

In PLANNER, if:

```
(THGOAL (GETON PERSON1 TRAIN))
```

is input, procedures 1–3 are applied in order to achieve the first sub-goal (AT PERSON1 STATION) of 1; and next, in order to achieve the second sub-goal (HAVE PERSON1 TICKET), 4 is activated. The first sub-goal of 4 is already satisfied, and therefore an attempt is next made to satisfy the second sub-goal (HAVE PERSON1 MONEY). When this is done, (AT PERSON1 STA-tion), which was achieved initially, becomes no longer true. This problem can be alleviated by changing the order of the sub-goals in 1. Even if, however, the problem is redone in this way, the procedure chain 3,2 derived so far must be discarded and the operation begun again from the beginning.

Given the reason for the failure of the initial trial, and that it can be overcome by adding 5 in front of 2, it would be convenient if such changes could be made. CONNIVER avoids the problems of PLANNER by giving the user flexible control of the search: functions are provided so that a procedure can be interrupted at any point to report a result to a higher procedure, and an interrupted procedure can be restarted from the same point by the higher procedure. In addition, points from which backtracking is possible may be specified, and backtracking can be carried out at any time. For this purpose, copies of the database must be kept for a large number of time-points, and in CONNIVER the problem state for each of these times is referred to as the *context.* How these contexts are handled is the next subject of discussion.

6.4.2 Context management

When a search is being carried out, the problem state must always be known. If the search strategy is fixed, either as depth-first or breadth-first then determining the state is relatively simple; but in CONNIVER the order in which the search will be carried out is not known, and the contexts are therefore saved in a tree structure. As an example, consider the problem of determining the next move in a game of chess. The given board position will be saved as context 0. We shall suppose that from that position a number of moves are possible and one of them puts the opponent in check. New contexts will be

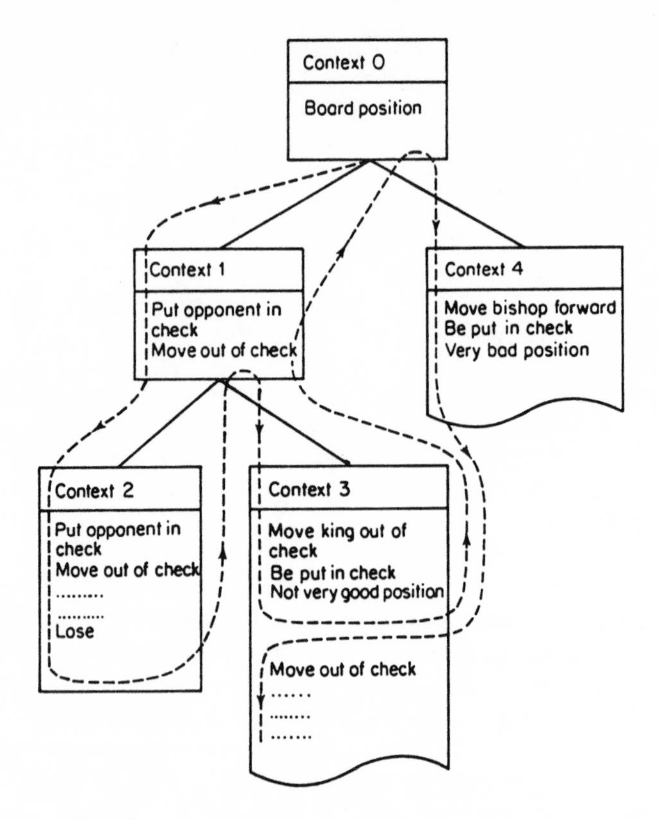

Fig. 64. Contexts and problem-solving control

generated as shown in Figure 64, and the changed position as a result of the check will be saved. Next, if the opponent moves his king out of check, context 1 will be written similarly. If at this point we wish to find out something about the board position, first we look at context 1, and if it is written there (for example the position of the opponent's king) we have the answer; whereas if it is not written there (for example the position of our king) we look at the next context up in the hierarchy. If we suppose that the move tree diverges here, then a new context 2 will be created, and a candidate for the next move written. Again, to find out something about the board position we look first at context 2, and, if it is not written there, at progressively higher contexts. Suppose, then, that carrying out a search in context 2 shows a lost position. In that case this context is no longer required and can be deleted. Next another move is considered, and the search proceeds with another context 3 created for that move. Suppose that at some stage the board position is considered not to be very good, and that another move is considered. In this case context 3 is saved as it is, and a context 4 for another move is created. If as a result of searching beyond that move it is determined that the board position is even worse than before, then that context is again saved as it is, and the search is recommenced after returning to context 3. In this way, whatever search is carried out there is always a correct record of the board position, without the required memory size being excessive.

As shown in this example, CONNIVER has contexts in a tree structure, to give the user free rein to control the problem solving. Since PLANNER is restricted to a depth-first search, if a procedure once fails there is no requirement to save the context at that point, and it may be deleted.

When a procedure is interrupted to execute another procedure, and then the interrupted procedure is to be resumed from after the interruption, it is necessary for the two procedures to call each other. Procedures which do this are called co-routines, and CONNIVER is provided with co-routine facilities.

6.5 PROLOG

PROLOG is a language for problem-solving based on predicate calculus. After procedural languages such as PLANNER and CONNIVER had been developed, the first proposals were made for regarding predicate calculus as a programming language. Here an overview of this technique will be given, using as a representative the predicate calculus language PROLOG, which was developed at Edinburgh University.

6.5.1 PROLOG format

The basic unit in PROLOG is called a *term*; the simplest terms, which are constants and variables, correspond to atoms in LISP. Constants are shown in lower-case letters and variables in capitals. Functions are also terms, as exemplified by the following:

point (X, Y, Z)
sentence (np(he),vp(v(likes),np(her)))

The latter item may be thought of as the tree structure of Fig. 65.

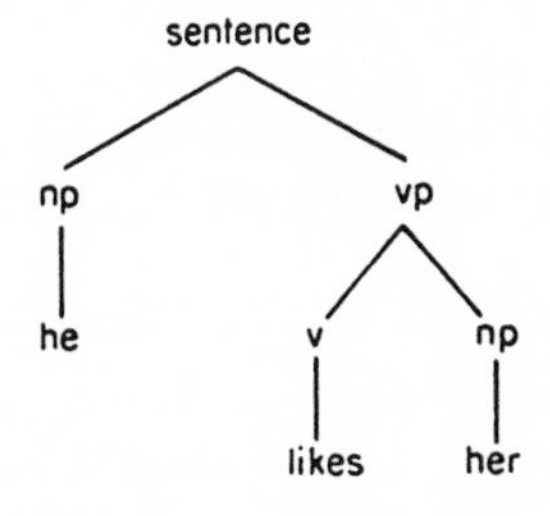

Fig. 65. Tree structure

Lists are defined similarly to those of LISP, but are enclosed in square brackets, with elements separated by commas. Thus (a b c) in LISP is expressed in PROLOG as [a,b,c]. By writing a list as:

```
[X,..L]
```

we indicate that the list CAR is X and the CDR is L; and:

```
[a,b,..L]
```

indicates that the list CAR is a, and the CDR is [b,..L].

A program is a sequence of statements, each of which consists of either or both a head and a body. The following statement, for example, contains both:

```
P:-Q,R,S.
```

P is the goal for the head and Q,R,S the goals for the body, and they are separated by ':-'. The meaning of this statement is 'If all of Q, R and S are true, then P is true'; or alternatively 'To establish goal P, establish Q, R and S'. If the head only is present, the format is:

```
P.
```

and this means 'P is true', or 'Goal P is achieved'. The representation when the body only is present is:

```
?-P,Q.
```

and this means 'Is $P \wedge Q$ true?', or 'Establish goals P and Q'.

P, Q and so forth can of course include variables. Note, though, that the head is always one term only, and that in the body a comma indicates a conjunction. To express a disjunction, we have already shown that it may be rewritten as a conjunction. Suppose, for example, we express 'For the grandfather of X to be Z, the mother of X must be Y or the father of X must be Y, and also the father of Y must be Z'; using a semicolon to indicate disjunction, then we have:

```
grandfather(X, Z):-(mother(X, Y);father(X, Y)),father(Y, Z).
```

This may be rewritten, however, as:

```
grandfather(X, Z):-mother(X, Y),father(Y, Z).
grandfather(X, Z):-father(X, Y),father(Y, Z).
```

Alternatively, by introducing a new function parent(X, Y), it may be expressed as follows:

```
grandfather(X, Z):-parent(X, Y),father(Y, Z).
parent(X, Y):-mother(X, Y).
parent(X, Y):-father(X, Y).
```

The principal characteristics of expressions in PROLOG have been described; as in LISP, a number of built-in functions are also provided, but discussion of them will be omitted here.

6.5.2 Basic operation of PROLOG

Now a predicate function descendant(X, Y) which is true if Y is a descendant of X will be defined as follows:

```
descendant(X, Y):-child(X, Y).
descendant(X, Z):-child(X, Y),descendant(Y, Z).
```

In other words 'Y is a descendant of X if Y is a child of X', and 'Z is a descendant of X if Y is a child of X and Z is a descendant of Y'. Suppose now that the following data is input:

```
child(alan, brian).
child(alan, lucy).
child(lucy, clive).
child(lucy, jane).
```

Note that the input statements consist of head portions only. Suppose now that the following question is asked:

```
?-descendant(alan, clive).
```

PROLOG will search for a statement having a head portion matching with the question, and the search is carried out from top to bottom. In this example, there is a match with the head of the first statement, simply by setting X=alan and Y=clive. When there is a head match, next an attempt is made to match the body of the statement. Since X and Y are determined, the body becomes child(alan, clive). This does not match with any statement, however. When there is a match failure, PROLOG automatically backtracks, deletes the immediately preceding successful match, and continues the search. Then the second statement is found. In this, X=alan and Z=clive, and the body becomes: child(alan, Y), descendant(Y, clive). The first term matches with the input statement child(alan, brian), and thus the second term becomes descendant(brian, clive). That match, however, fails. Therefore, once again the system backtracks, and this time the first term matches with child(alan, lucy). The second term then becomes descendant(lucy, clive), which matches with the head of the first statement, so that the body of the first statement becomes child(lucy, clive), which matches with the input. The answer is therefore 'true' (yes).

Thus the search strategy is depth-first as in PLANNER. Moreover, when there is a match between a goal statement and the body of a statement, variable unification is carried out in the same way as in the resolution principle.

We shall now consider another example:

```
?-descendant(alan, X).
```

If the search is carried out in the same way as before, then first there is a match with the first statement, the body of that statement child(alan, X) matches the data child(alan, brian), and thus X=brian. With this the search succeeds. If this X fails to meet some other condition (if when the answer is returned, ';' is input, that answer is regarded as failing), then PROLOG recontinues the search. Thus, next X=lucy is obtained, and then successively clive and jane. If backtracking is then forced again, the search fails and returns 'false' (no). Thus:

```
X=brian;
X=lucy;
X=clive;
X=jane;
no
```

In addition to calculating the value of a function, the system finds values of variables which give a true value.

When it is not desired to carry out automatic backtracking, the cut operator '!' can be used. Suppose, for example, that a list of the personnel in a company department is ordered by age, and that each person's age is given as age(X). A function corresponding to MEMBER in LISP may be defined as follows:

```
member(X, [X,..L]).
member(X, [Y,..L]):-member(X, L).
```

Suppose we wish to find people not over 40; then we input:

```
?-member(X, [smith, jones, roberts, davis])!,   age(X)=<40.
```

(Operators in PROLOG include =, =< etc.) The first term above matches member(X, [X,..L]), and thus X=smith. Next the system tests whether age(smith)=<40. Even if this does not hold, because of the presence of the '!' there is no backtracking, and no other value for X is sought. Thus the exclamation mark is used in cases where there is no point in further continuing the search.

Next a function corresponding to APPEND in LISP will be defined, where append(L1, L2, L3) means that the list L3 is that formed by concatenating lists L1 and L2:

```
append([X,..L1], L2, [X,..L3]):-append(L1, L2, L3).
append([], L, L).
```

Now suppose that the following is input:

```
?-append([a], [b], X).
```

This matches with the head of the first statement in the definition of append. To simplify, we first replace X in the first statement by Y, and then we have: Y=a, L1=[], L2=[b], X=[a,..L3]. The body then becomes append([],[b], L3); and comparing with the second statement, by replacement this gives L=[b], L3=[b]. From this it will be seen that the variable X being sought is [a, b]. Since in general the definition of append may be used in comparisons many times, a predetermined variable name like L3 cannot be used. Instead, the system generates a new variable each time it is called. For simplicity these variables will here be taken as X1, X2, etc. Then the above problem solution may be expressed as follows:

```
append([a], [b], X)
(1) append([],[b], X1), x=[a,..X1]
(2) X1=[b], X=[a,..X1]
(3) X=[a,b]
```

Similarly:

```
append([a, b], [c], X)
(1) append([a,..[b]], [c], X)
(2) append([b], [c], X1), X=[a,..X1]
(3) append([], [c], X2), X1=[b,..X2], X=[a,..X1]
(4) X2=[c], X1=[b,..X2], X=[a,..X1]
(5) X1=[b, c], X=[a,..X1]
(6) X=[a,b,c]
```

Next, if the input is:

```
?-append(X, Y, [a,b,c]).
```

then if backtracking is forced, the following sequence of replies is obtained:

```
X=[a,b,c]
Y=[];
X=[a,b]
Y=[c];
X=[a]
Y=[b,c];
X=[]
Y=[a,b,c];
no
```

For an input of:

```
?-append(X, Y, [a,b,c]), length(X, 2).
```

the reply will be:

X=[a,b]
Y=[c]

(Here length(X, 2) is a predicate function which means that the length of list X is 2.)

6.5.3 PROLOG characteristics

It will be seen from its basic operation that PROLOG has many points of similarity with PLANNER. Specifically:

(1) Facts and procedures are retrieved by pattern-matching. PROLOG differs from PLANNER in that the order of attempting matches follows the input order of statements. The user can therefore control the execution order of the program.
(2) Searching is carried out with automatic backtracking, but the cut operator '!' may be used to block the search.

The following are points of comparison with conventional programming languages:

(1) List processing does not depend on CAR and CDR as in LISP, but is based on pattern matching.
(2) Functions are based on predicate functions, and there is no distinction between input variables and output variables. To find the length of a list in LISP the following is used:

    (SETQ M (LENGTH L))

    In PROLOG, we use the predicate function:

    length(L, M)

    From this function, in addition to evaluating true or false for input values of L and M, it is possible to treat L or M as an unknown.
(3) The format of input statement is restricted. The basic PROLOG statement is:

    P:-Q,R,S.

    The head is a single term, and the body is a conjunction of terms not including negations. Expressed in logic terms this becomes:

    $Q \wedge R \wedge S \Rightarrow P$ or $\sim Q \vee \sim R \vee \sim S \vee P$

    Thus when expressed as a clause, there is at most one positive literal. A set of logical formulae of this type is termed a *Horn* set, and it may therefore be said that PROLOG applies the resolution principle to Horn sets.

It will be seen, then, that by placing some restrictions on the input statements and by carrying out matching in the order that the input statements are given, PROLOG leaves the resolution strategy to the user. The system is therefore compact, and faster execution can be provided.

Chapter 7

Representation and Use of Knowledge

It goes without saying that knowledge is required to solve a problem, but it is not clear just what is meant by knowledge in this case. At the beginning of this book, solving a problem was defined as determining a suitable sequence of operators. This means that the knowledge required to solve a problem consists of a set of operators and a strategy for determining a suitable sequence. In Chapter 5 it was shown that facts and possible inferences can be expressed by first-order predicate calculus. In that case the knowledge consists of the propositions expressed in first-order predicate calculus and the strategy for proof based on the resolution principle. Thus knowledge includes facts relating to the object world, the rules (operators or possible inferences) needed to produce new facts (the problem state), and a strategy for rule application. In first-order predicate calculus facts and rules were not distinguished, and were handled uniformly as propositions; it is sometimes possible to handle knowledge in this way without classification.

Up to Chapter 5 the discussion centred on strategies for rule application, and we considered how a relatively small number of rules and facts could be combined so that the problem was solvable. In this chapter, looking at problem-solving from another angle, we discuss how facts and rules should be expressed to assist the solution. From here onwards, what will be termed simply 'knowledge' is facts and rules, and problem-solving strategies are excluded. Not only problems with a toy world, such as the building block problem, but also problems requiring a large volume of knowledge, such as understanding human speech or giving a medical diagnosis, will be considered.

7.1 PROCEDURAL KNOWLEDGE AND DECLARATIVE KNOWLEDGE

The terms 'procedural knowledge' and 'declarative knowledge' are used contrastingly. A typical example of procedural knowledge is a conventional computer program, where the entire problem-solving procedure is written as a program. For example, a procedure to make a list by joining together two lists L1 and L2 may be written in LISP as follows:

```
(DEFUN APPEND(L1 L2)
  (COND ((NULL L1) L2)
        (T (CONS (CAR L1) (APPEND (CDR L1) L2)))))
```

If the same procedure is written declaratively, using the language PROLOG described in the previous chapter, it will be thus:

```
append([X,..L1], L2, [X,..L3]) :- append(L1, L2, L3).
append([ ], L, L).
```

Seeking a response by the LISP procedure is more efficient than by using this declarative description and getting the response by pattern-matching. Additionally, there is a large set of procedures, for sorting a large number of numbers, finding mean values and distribution and so forth, which are best expressed as procedural knowledge.

One procedure, however, is restricted to one processing objective. The function APPEND always takes two arguments, and returns the list made by joining them. PROLOG, on the other hand, is somewhat more universal, and as shown in the previous chapter, using its notation responses may be obtained to a number of different queries. We show an example again:

```
? - append([a, b], [c], X).
X = [a, b, c]
? - append([a, b], [c], [a, b, c]).
yes
? - append(X, [c], [a, b, c]).
X=[a, b]
? - append(X, Y, [a, b, c]).
X = [a, b, c]
Y = [ ]
```

The expression of knowledge by predicate logic is the archetype of declarative knowledge. Assume, for example, that the following proposition is declared:

$$\sim P(x, y) \vee Q(x, y)$$

This has many interpretations, such as 'If $P(x, y)$ then $Q(x, y)$', 'If $\sim Q(x, y)$ then $\sim P(x, y)$', or 'If $P(a, f(a))$ then $Q(z, f(a))$', and different interpretations may be used for different purposes. The rewriting rules discussed in Chapter 4 are also an example of declarative knowledge.

PLANNER is an example of a language in which, in order that a procedure may be used for a larger number of purposes, a procedure call is defined by a pattern, and procedures are activated automatically. Since the number of

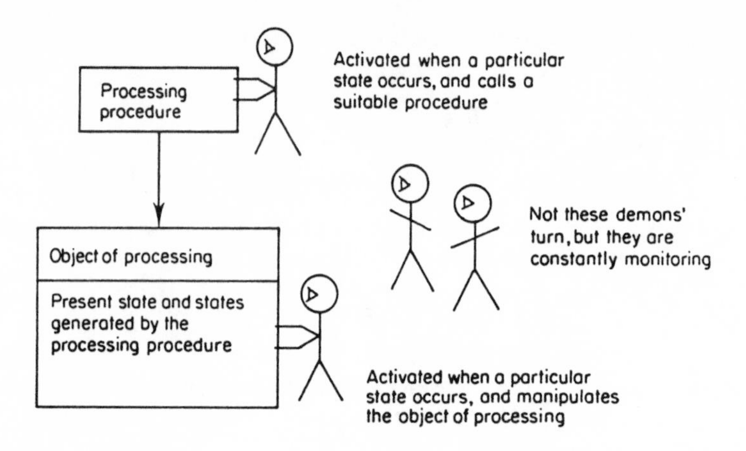

Fig. 66. Demons: a typical situation

procedures matching the pattern is not limited to one, a search is necessary, and PLANNER carries out a depth-first search. In this case, though, the benefit of having procedures is much reduced, and efficiency is lost. As a result systems such as CONNIVER, in which the search may be controlled at will, have been proposed. In this case, on the other hand, the universality of the procedures is once again impaired. Thus it will be seen that both procedural knowledge and declarative knowledge have their advantages, and just where the compromise may be found will depend on the particular problem under consideration.

7.2 DEMONS

Knowledge is normally used by being called by other knowledge. In a conventional program, a procedure calls another procedure by name, whereas in PLANNER procedures are called by a pattern. Fig. 66 shows a typical example of the concept of 'demon'; these constantly monitor the situation, and when a predefined circumstance occurs, they jump out and carry out the appropriate action. In other words, when a demon will be activated is not defined by some other activating entity, but by the demon itself.

The demon concept is used everywhere. In the operation of a conventional computer, for example, when a power irregularity, an I/O error or a program error occurs, the reason for the failure is stored at some memory location or is output, before stopping the program. This may be regarded as being done by a demon constantly monitoring for an error state.

The previous chapter described how PLANNER procedures include a consequent type procedure THCONSE activated when necessary, an antecedent type procedure THANTE activated when a particular fact is added to the database, and a deletion procedure THERASING activated when a particular fact is deleted from the database. Antecedent and deletion procedures are examples of demons.

7.2.1 Application to natural language comprehension

Natural language comprehension was the first field in which the efficacy of demons as a method of problem-solving was shown. It was shown that in order to understand an English story it is not sufficient to understand the meaning of each individual sentence.

The following story (an extract) will be used to show the operation of demons:

Mary has been invited to John's party.
She thought he might like a kite.
She went to her room and shook her money-box.
Taking the 50 pence that came out, she went to the toy shop.

Now suppose we extract the meaning from the four sentences making up this story. For simplicity, verb tenses will be ignored. Thus the meaning derived by analysing the sentences may be represented as follows:

(Mary is-invited (John's party))
(Mary think (John like kite))
(Mary go (Mary's room))
(Mary shake money-box)
(Mary take 50-pence)
(Mary go toy-shop)

Can the story be said to have been understood by means of this semantic expression? Could questions like 'Why did Mary think John would like a kite?' or 'Why did she shake her money box', which are self-evident to a human, be answered?

To understand the story properly requires the use of knowledge not written in the story itself. In this example the following are only a few examples of common-sense knowledge required:

If invited to a birthday party, take a present.
To take a present, it is sufficient to buy one.
To buy a present, money is required.
If a money box is shaken upside down, money will come out.

This kind of common-sense knowledge is not written in an ordinary story. The problem here is how to represent and use common-sense knowledge in a computer, and demons have been proposed as a possible approach.

Demons are expressed as follows:

```
(DEMON demon name
  (variable list)
  (monitoring pattern)
  (GOAL (conditions for demon to be activated)
          ...
  (action when activated)
          ...  )
```

A demon for the birthday party example may be expressed as follows:

```
(DEMON birthday-party
  (X Y)
  (?X   is-invited (?Y's birthday-party))
  (ASSERT (?X consider (?Y's present)))
  (ASSERT (?X consider (birthday-party-going clothes))));
```

if a pattern matching (?X is-invited (?Y's birthday-party)) appears in the database, add the propositions following ASSERT to the database.

Note that variables which are to be matched with anything are prefixed by '?', and also that in this example GOAL is not used. The birthday party demon expresses the common-sense idea: 'If invited to a birthday party, think what present to give, and what to wear'.

Let us define some more demons:

```
(DEMON give-a-present
  (X Y Z)
  (?X think (?Y like ?Z))
  (GOAL (?X consider (?Y's present)))
  (ASSERT (?X give ?Y ?Z)))
```

; if the pattern 'X thinks Y likes Z' appears, the proposition following GOAL is inspected (X is considering a present to Y) and if this proposition holds, the fact that X intends to give Z to Y is added.

```
(DEMON buy-a-present
  (X Y Z)
  (OR (?X want money)
       (moneybox))
  (GOAL (?X give ?Y ?Z))
  (ASSERT (?X buy ?Z))
  (ASSERT (?X need money))
```

; either of the two patterns following OR is sufficient to start.

```
(DEMON extract-money-from-moneybox
  (X)
  (OR (?X open lid)
```

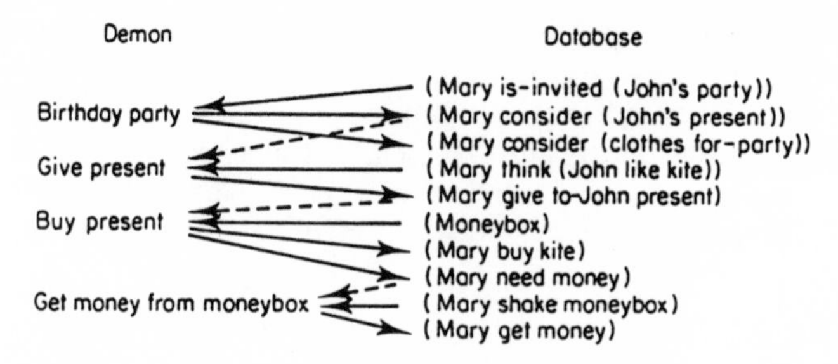

Fig. 67. Example of demon operation. Solid lines show monitoring patterns, broken lines are activating conditions

```
        (?X shake moneybox)
        (?X break moneybox))
    (GOAL (?X require money))
    (ASSERT (?X get money))).
```

We shall now investigate how each of these demons works, referring to Fig. 67. If the first sentence is input, its semantic expression may be obtained. Next the birthday party demon is activated, two inferences are made, and these are added to the database. Then the next sentence is input, and its semantic expression obtained. This matches the pattern monitored by the present giving demon, and so that demon's starting conditions are checked. The starting conditions are satisfied, and the inference that Mary wants to give a present to John is made. In this way various demons are activated, and interpretation of the story proceeds. If the interpretation is carried out as in Fig. 67, answering the questions given earlier will be simple. For example, the question 'Why did Mary shake her moneybox?' may be answered by tracing the arrows in Fig. 67: 'To get money out'. Similarly, in response to 'Why did she shake her moneybox?', tracing the arrow backwards from 'Get money out of moneybox' the answer may be given: 'Because she needed money'.

7.2.2 Demon management

There are various methods of implementing demons in a computer system, but the following is the method most generally used. A list of the demons is made, and then as the processing proceeds, each time the state changes, the changes are made into a list, and for each demon the monitoring pattern is checked. If there is a match, control is immediately passed to the demon. If the state is not changed by the action of the demon the check continues; but if there is a change, then because of possible new changes, the check must begin again from the beginning. This method can handle any kind of demon, but the time required for processing is large.

One method of using demons efficiently is to generate them when required, and delete them when no longer needed. In the example above, instead of activating all the demons from the beginning, by some method or other only the birthday party demon is activated. When the birthday party demon has started, the give-a-present demon may be added. Thus, within the operation of a demon will be included demon activation and deletion, and this will be written as follows:

(PUTIN demon-name) generate demon
(PUTAWAY demon-name); delete demon

Demon management operations will be added to the operation of the demons in the example described above, as shown in Table 11.

TABLE 11

Demon name	*Added operations*
Birthday party	(PUTIN give-a-present) (PUTIN choose-clothes)
Give-a-present	(PUTIN make-a-present) (PUTIN buy-a-present)
Buy-a-present	(PUTIN get-money-from-moneybox) (PUTIN shopping) (PUTAWAY make-a-present)

Now if we also assume that after a demon is activated it is deleted, then always only demons which are likely to prove useful will be monitoring the situation. In this example, at the point where the fourth sentence has been processed, the choose-clothes and shopping demons remain. The shopping demon may help in interpretation of the last sentence. The choose-clothes demon may be called later if the story continues. This example shows clearly that in some cases a demon will disappear without even being called once.

General criteria have not yet been established as to what demons should be provided, or how demon management should be carried out in order to understand a large number of stories.

7.3 PRODUCTION SYSTEMS

The demons described in the previous section were used in combination with other processing procedures, but the production systems in this section describe all of the processing just by sets of demons.

Production systems were proposed as a framework for human thought, but using this way of thinking has developed into research to attempt to solve complicated problems. After a basic explanation of production systems, this application will be considered.

7.3.1 Basic operation of production systems

A simple example will serve to show how production systems work. A sentence will be subjected to syntax analysis by means of the grammar used in Section 4.1. In a production system, knowledge is expressed as a collection of rules. In this example the grammar is expressed as a collection of rewriting rules, which will be called production rules (simply 'rules' where this is unambiguous):

R1 SUB PRED → S
R2 PRON → SUB
R3 NP → SUB
R4 V NP → PRED
R5 DET N → NP
R6 he → PRON
R7 saw → V
R8 a → DET
R9 dog → N
R10 cat → N

Now suppose that the following input sentence is given:

He saw a dog. (25)

First, store this sentence temporarily, in what will be termed *short-term* memory or *working* memory. The production system compares the contents of the working memory with the left-hand sides of the production rules, and where there is a match, replaces the matching portion with the right-hand side of the rule. For (25), the left-hand sides of rules 6–9 all match, and the working memory is rewritten:

PRON V DET N

Now rules 2 and 5 match, and as a result the working memory becomes:

SUB V NP

and finally, by rule 4 and rule 1 becomes S. Thus a syntactical analysis of the English sentence has been carried out; the order of application of the rules need not necessarily be the same.

Consider, on the other hand, the following:

Saw he a dog cat. (26)

If the rules are applied in the same way as before, the sequence in Fig. 68 will be obtained. Since S is not finally derived, it is deduced that this is not a sentence; a

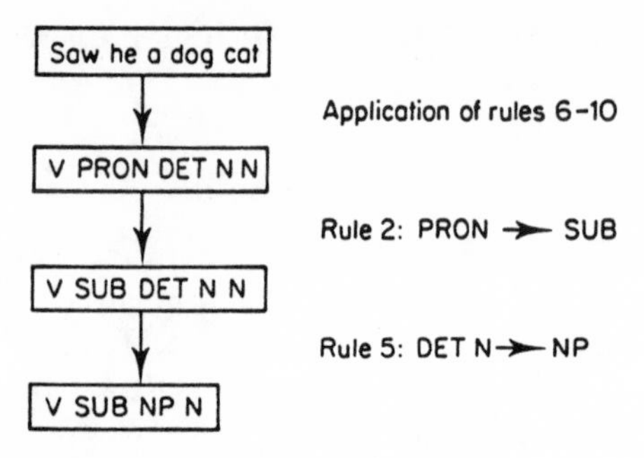

Fig. 68. Sentence recognition by a production system (example of failure)

check should be made, however, that even if the order of application of the rules is changed S cannot be obtained.

7.3.2 Production system model

A production system may be thought of as consisting of the following elements:

(1) a set of production rules
(2) a working memory
(3) some means of control for determining the order of application of the rules.

The rules generally take the following form:

Precondition $C_1 \wedge C_2 \wedge \ldots \wedge C_m$
Action A

The precondition is a conjunction of a number of predicates, and is shown:

$$C_1 \wedge C_2 \wedge \ldots \wedge C_m \rightarrow A$$

This also corresponds exactly to the PROLOG expression:

$$A\text{:- } C_1, C_2, \ldots, C_m$$

(except that in the production rules C_i can be a negation).

Fig. 69 shows a general framework for a production system. It will be seen that the working memory is changed not only by the application of the rules, but also by changes in the outside world. It can be argued that human or animal behaviour is modelled in this way. In other words, a large number of rules are stored in a long-term memory, and behaviour is then governed by a production system with a relatively small working memory. In this interpretation, the process is a repeated alternation of the following two steps:

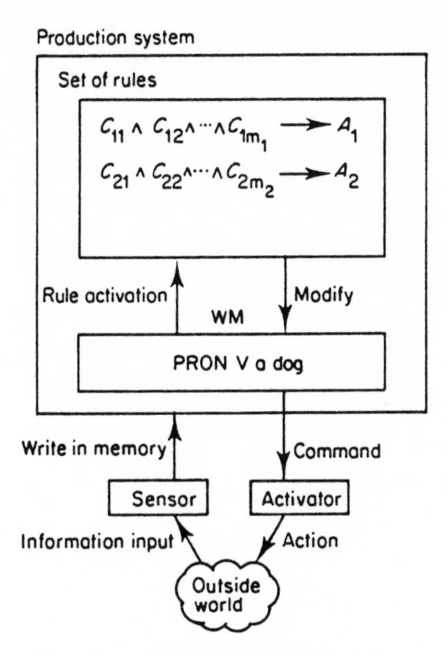

Fig. 69. Overall structure of a production system

Step 1: Comparison of state of working memory with rule preconditions
Step 2: Application of rules for which a match was achieved

This repetition is known as the *recognition–act* cycle.

There may, however, be more than one rule for which a match is achieved in step 1. In this case, the outcome may vary depending on the rule which is activated; it is the function of the third element of the production system to control what happens in such situations. The method of doing this will be described later.

7.3.3 Comparison of production systems with procedural knowledge

For the purposes of comparison, consider a procedure to carry out the grammatical analysis described earlier. The syntax rules are shown in Fig. 70 as a transition network: the circles correspond to states, and the arrows to transitions. For example, at the bottom left of the diagram, in state N if 'dog' or 'cat' is found, a transition to state N_f occurs. The suffix f indicates that processing has completed successfully. When an arrow is labelled 'call SUB', it indicates a jump to state SUB, and when that process has completed successfully, a return to the original arrow and a transition to the next state. For example, in the second line of the figure, in state SUB, a jump is made to PRON, and if there 'he' is found, a transition is made to state SUB_f. If a string

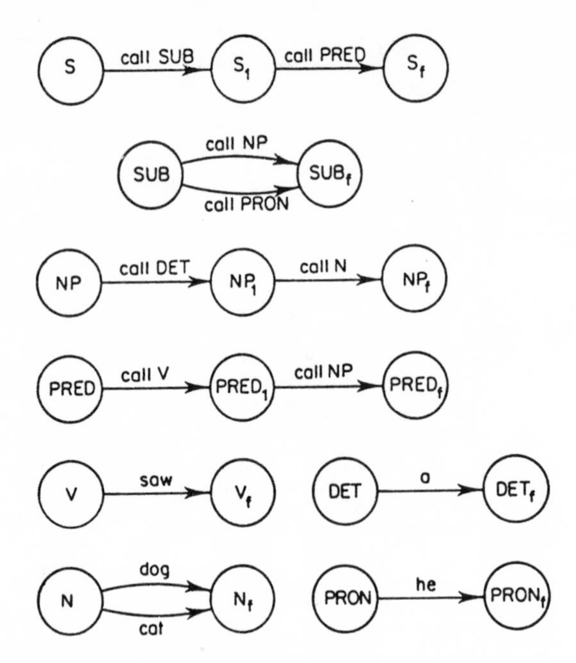

Fig. 70. Syntax expressed as a transition network

of words is supplied, and starting from state S, state S_f is reached, then provided that all of the string has been used, the input has been determined to be a sentence.

This transition network can be expressed as a program. The input word string is taken to be a list of atoms:

```
(SETQ L '(HE SAW A DOG))
```

First functions to process the terminal symbols are defined:

```
(DEFUN PRON
  (COND((EQUAL(CAR L) 'HE) (SETQ L(CDR L)))))
```

The function V is the same as PRON, but with HE replaced by SAW:

```
(DEFUN DET
  (EQUAL(CAR L) 'A))
(DEFUN N(L)
  (OR (EQUAL(CAR L) 'DOG)
       (EQUAL(CAR L) 'CAT)))
```

Next, functions for processing non-terminal symbols are defined:

```
(DEFUN S
```

```
  (COND (SUB (COND (PRED (NULL L))))))
; S succeeds only when SUB and PRED succeed and as a result L has
become NIL.
(DEFUN SUB
  (OR NP PRON))
(DEFUN NP
  (COND (DET (COND ((N(CDR L))
                    (SETQ L (CDR (CDR L))))))))
; L is replaced only when DET and N have both succeeded.
(DEFUN PRED
  (COND(V NP)))
```

Notice particularly that, in the above program, the processing sequence is specified. In other words, if (S L) is input, then first SUB is evaluated, and if it succeeds, PRED is evaluated. In SUB, NP is first evaluated, and if this fails, PRON is evaluated.

The order of application of the rules of a production system, on the other hand, is determined by the state of the working memory at execution time; the rules themselves are mutually independent. Therefore updates to add or remove rules are easily done.

Suppose, for example, that the following new rule is to be added:

R11 DET ADJ N $\rightarrow$ NP

In a production system, it is sufficient to add rule 11 to the existing set of rules. If the pattern DET ADJ N occurs in the working memory, then rule 11 will be called, and this will be replaced by NP. For the program, on the other hand, a decision must be made as to where to insert the new rule, and the program amended accordingly. In this case, the function ADJ is added, and NP is amended as follows:

```
(DEFUN NP
  (COND (DET (COND ((N(CDR L))
                   (SETQ L (CDR (CDR L))))
                   ((ADJ (CDR L))
                   (COND (N (CDR (CDR L)))
                         (SETQ L (CDR (CDR (CDR
                         L)))))))))))
```

As the number of rules increases, so does the number of amendments needed when one rule changes.

This problem does not exist in a production system; but since it must be determined at execution time which rules should be applied, a large amount of processing time is required.

7.3.4 Choice of rules to apply

When comparing the state of the working memory with the preconditions of the rules to find a rule which can be applied, more than one rule will sometimes be found. In some cases the result will be the same whichever rule is applied, but generally it will vary. Selecting one rule from those which are applicable is termed *conflict resolution*. This is the same problem as selecting an operator in a search. In a production system, the following selection methods might be considered:

(1) Form the rules into a list, and apply the first matching rule. In other words, attach a priority ranking to the rules.
(2) Give priority to the rule with the most strict precondition. Since each precondition is written as a combination of clauses, this would give priority to the precondition with the largest number of clauses. In the above example of syntax analysis this gives priority to the rules with the longest left-hand sides, and this strategy is also known as *longest matching*.
(3) Give priority to the most recently used rule.
(4) Give priority to the rule concerned with the most recently used variable.
(5) Give priority to the rule most recently added to the set of rules.
(6) At execution time compute a priority value for each rule, and select the rule with the highest priority value.

Even if rule selection is based on a heuristic standard as above, the correct result will not always necessarily be derived. In the case of a search, the problem state is stored in memory, and in difficult cases other possibilities are investigated. If the same type of search is attempted in a production system, then a number of working memories must be maintained. Methods including this kind of search are not normally, however, included in the narrow definition of a production system.

In syntax analysis, for example, there will be two rules depending on whether 'that' is classified as a pronoun or as a relative pronoun; the results will be different depending on which rule is applied. A simple production system cannot, therefore, be used for syntax analysis.

7.3.5 Forwards deduction systems based on production rules

As is the case with Newton's laws in mechanics, a small number of rules can sometimes solve a large number of problems; but in other cases, such as medical diagnosis, the solution must be found through a large body of empirical knowledge. Problem-solving systems which are provided with this type of knowledge to the same level as a trained expert are called *expert systems*, and the study of such methods is called *knowledge engineering*.

There are still not many examples of the application of knowledge engineering, but the principal systems are based on the concepts of production systems. A typical example in DENDRAL, which predicts the structural formula of an organic chemical compound from its qualitative analysis data. Finding the chemical formula of the compound is easy, but determining its structural formula requires specialist knowledge.

DENDRAL has three parts. The first is the structure generating part, which generates possible structural formulae from the chemical formula. The second part is a spectrum predictor which derives from a structural formula the corresponding predicted mass spectrum, and compares this with the actual data. These two components are in principle sufficient; it is enough to predict the spectrum from a structure generated by the structure generator, and test whether it agrees with the data. Unfortunately, however, the structure generator substance represented by $C_nH_{(2n+2)}O$, there are seven possible structures for $n = 4$, and about 1000 possibilities for $n = 10$. $C_6H_{13}NO_2$ has about 10 000 possible structures. It is, therefore, inefficient to investigate all the possibilities.

This is where the third part is required. It is a structure predictor, which deduces partial structures from the data and predicts impossible structures. This part corresponds to the knowledge of a human expert.

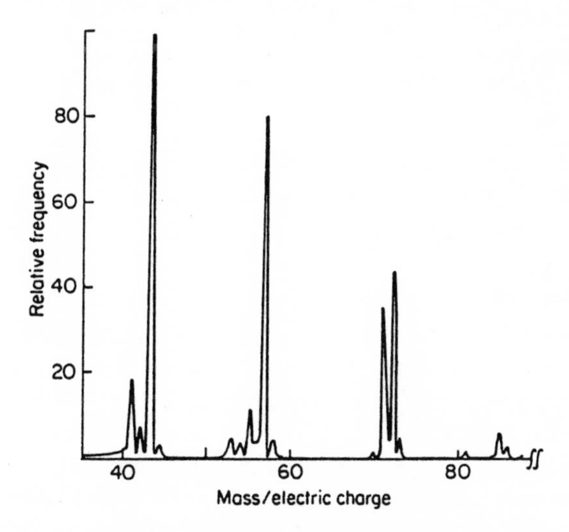

Fig. 71. Portion of mass spectrum

The original data is a spectrum as shown in Figure 71, which is digitized and input in the following form:

((peak 1 position) (peak 1 value) (peak 2 position) (peak 2 value)
... ... (peak m position) (peak m value))

In this example:

((41 18) (42 7) (43 100) (44 3) (53 3) (54 1)...)

This is a spectrum for $C_8H_{16}O$, which has several thousand candidate structures. Only about 700 remain, however, if chemically unstable structures are eliminated.

The structure predictor is formed from production rules made by asking an expert. These rules are expressed as follows:

Rule 1
If: there is a high peak at the point where mass/charge is 71,
there is a high peak at the point where mass/charge is 43,
there is a high peak at the point where mass/charge is 86,
and there is some peak at the point where mass/charge is 58,
then: the partial structure N-propylketone-3 may be assumed to be present.

Rule 2
If: there is a high peak at the point where mass/charge is 71,
there is a high peak at the point where mass/charge is 43,
there is a high peak at the point where mass/charge is 86,
and there is no peak at the point where mass/charge is 58,
then: the partial structure isopropylketone-3 may be assumed to be present.

There are many such rules for ketones, ethers, alcohols and so on. From the spectrum for $C_8H_{16}O$ in Fig. 71, the partial structure N-propylketone-3 predicted. This structure is:

```
    H H H O H   H
    | | | || |  | |
  H-C-C-C-C-C-C-C-
    | | |    | | |
    H H H    H
```

If this structure is taken to be correct, then it is only necessary to join to it one carbon and six hydrogen atoms. The number of ways of adding these is strictly limited; and in fact when the rules are applied, there are about 40 candidate ketone structures.

For each candidate, the spectrum predictor deduces a spectrum. This part also operates by using production rules. Each rule is an expression of what spectral data would be obtained if a particular partial structure were analysed. When candidates are selected on the basis of the data being close to the predicted spectrum, two or three structures are obtained; in this example only one structure remains, and this is 4-octanone, with the following structure:

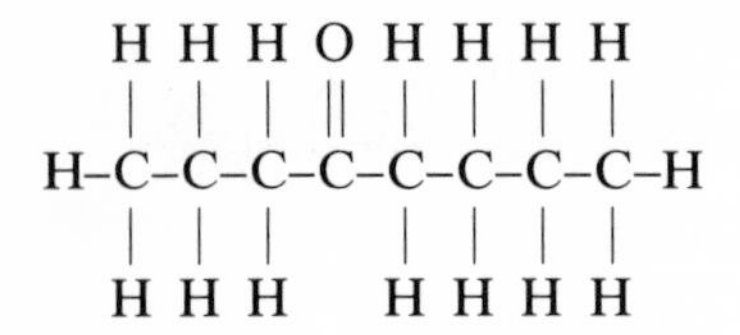

DENDRAL was developed by the Computer Science Department and the Mass Spectrometry Group at Stanford University, and is at present being used in the university's chemical research.

This system starts from data, and derives a conclusion from the forwards application of production rules. Systems applying production rules also exist in which processing is carried out in the reverse direction, and these will be described next.

7.3.6 Backward deduction systems based on production rules

Production rules may also be applied in reverse to solve a problem. Backward deduction systems postulate a conclusion and then determine whether or not that conclusion is true. The basic operation will be explained, taking as an example a system which makes medical diagnoses. First we hypothesize the name of an illness to be the diagnosis, and then investigate whether there are any contradictions with the patient's symptoms.

As described in Section 7.3.2, a production rule has the form:

$$C_1 \wedge C_2 \wedge \ldots \wedge C_m \rightarrow A$$

If the name of an illness is postulated, then a rule which provides that illness as a conclusion is sought. If a rule is found, then we investigate whether the precondition of that rule is fulfilled; all of the predicates C_i forming the precondition must be true. We investigate whether each C_i agrees with the symptoms; and if it cannot be directly derived from the symptoms, we seek a rule which has C_i as its conclusion. By repeating this step it can be determined whether the initial hypothesis holds. If it does not hold then another hypothesis is adopted.

In practice not every production rule will necessarily be true. For example, even if there is a rule 'If you have a temperature and a cough, and feel shivery, you have a cold', that will be true in most cases, but there will be exceptions. In this case an assessment value may be introduced, corresponding to the probability that the rule is true.

For infectious diseases of the blood, the expert system MYCIN, which hypothesizes which microorganism is causing the disease and advises on a drug for its treatment, assigns a certainty factor to each rule. The rules then have the following form:

Rule
If: the infection is a primary bacteraemia,
    the culture medium is a sterile site,
    the entry of the bacillus is thought to be through the gastrointestinal tract,
then: there is an indication that the bacillus is a bacteroid (0.7).

The final 0.7 is a certainty factor, which is a number which may range from 0 to 1, where 1 indicates complete certainty.

The facts on which the reasoning depends may also not consist entirely of certainties. In a short time interval no definite conclusion can be reached as to whether the bacillus being cultured is aerobic or not; thus each fact has a certainty factor attached to it. The facts are shown as follows:

The culture medium is blood (1.0).
The cultured bacillus is aerobic (0.25).
The cultured bacillus is *Escherichia coli* (0.75).

Backwards deduction reverts to an AND/OR tree search. Suppose that there are two rules with A as conclusion:

$$C_{11} \wedge C_{12} \wedge C_{13} \rightarrow A \ (CF_1)$$
$$C_{21} \wedge C_{22} \rightarrow A \ (CF_2)$$

Here, CF_i are the certainty factors. An AND/OR tree for these rules is as shown in Fig. 72. If we suppose A and trace the AND/OR tree downwards, we can determine the degree of certainty with which A holds.

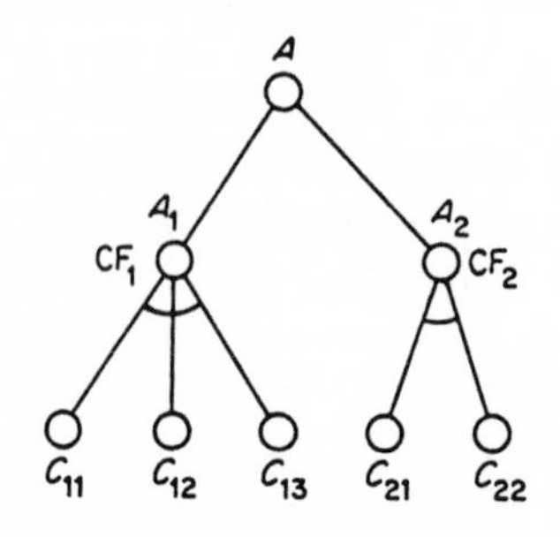

Fig. 72. Expression of production rules as an AND/OR tree

If the certainty, or probability of correctness, of C_{ij} is shown by $f(C_{ij})$, then the certainty of the conclusion A_1 (= A) derived from the first rule is:

$$f(A_1) = CF_1 \times \min\{f(C_{11}), f(C_{12}, f(C_{13})\}$$

Next a method of deriving $f(A)$ from $f(A_1)$ and $f(A_2)$ is required. According to the result of Chapter 4, this is:

$$f(A) = \max\{f(A_1), f(A_2)\}$$

In MYCIN, however, the following expression is adopted:

$$f(A) = f(A_1) + (1 - f(A_1)) \times f(A_2)$$

In other words, if there are many rules with the conclusion A, each makes a contribution to $f(A)$. If $f(A)$ is 0.2 or less, A is regarded as not holding, and $f(A)$ is taken as zero.

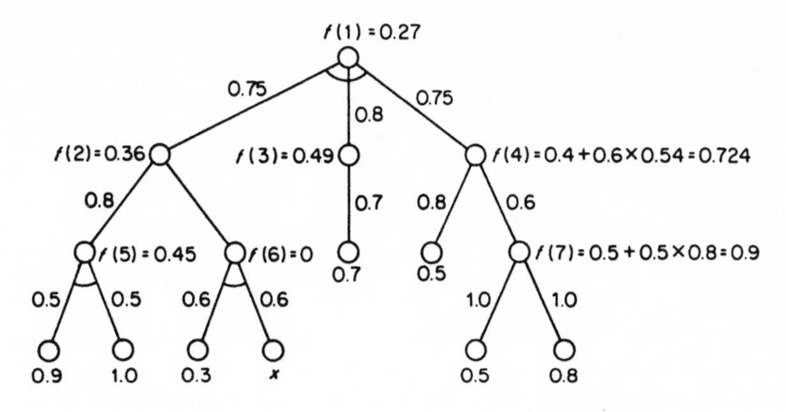

Fig. 73. Evaluation of certainties for an AND/OR tree

Fig. 73 shows an example of the steps in determining the certainty. The numbers beside the branches of the AND/OR tree are certainty factors. The terminal points correspond to the data, to which certainty factors are attached by the user (the doctor). In the fourth position from the bottom left only x appears; but this indicates that since whatever the value of x, $f(6) = 0$, x need not be known. In other words, once it is realized that $f(n)$ will be 0.2 or less, the node n need not be further investigated.

MYCIN collects data initially by asking ordinary questions of the user. Then it postulates the pathogen, and calculates its certainty, interrogating the user for any information necessary for the calculation. If the certainty value is 0.2 or less, then it makes another hypothesis and carries out the same procedure. Requests for results of time-consuming tests are delayed as far as possible, and only made when absolutely necessary. If after this several candidates for the pathogen emerge, an antibiotic effective against all of them is recommended. This part operates by forwards deduction using production rules.

The object of MYCIN is as far as possible to pin down the cause of an illness, and avoid the indiscriminate use of unnecessary antibiotics. Its success has been a stimulus to the development of many expert systems.

7.3.7 Features of deduction systems based on production rules

The following are characteristics of writing a large-scale system for practical use using production rules:

(1) Adding or amending knowledge is straightforward.
(2) It is easy to find out what knowledge is contained in the system.
(3) Processing can be flexible, depending on the input data.

(4) By tracing the rules used in the stages of deriving the conclusion, the user can understand why the system reached a particular conclusion. If the reasoning cannot be accepted, the rules may be changed.

(5) Since it is necessary not only to apply rules, but also to determine which rules should be applied, the processing speed is reduced.

(6) From a set of rules the system behaviour cannot be predicted.

(7) A procedure sequence cannot be directly represented.

From this list it will be seen that there are both advantages and disadvantages, but the majority of expert systems are at present based on production rules. Depending on the field of application, it may be better, rather than subdividing knowledge into production rules, to supply knowledge collected into chunks for each topic. This method of expression is the subject of the next section, and has, moreover, been incorporated into some expert systems.

7.4 HUMAN STRUCTURING

Humans carry out everyday thinking in an efficient manner. Even if a computer were provided with logical expressions corresponding to human common-sense knowledge, it would be difficult to extract conclusions efficiently in the same way as humans. It appears that in order to provide power and speed comparable to human thought, the units of knowledge must be bigger, and knowledge must be structured in chunks.

This approach sprang from attempts in psychological research to explain human problem-solving, and from research into natural language understanding. M. Minsky at MIT produced from these studies a 'frame theory', which simply gave general guidelines for knowledge representation, but which was soon the basis for several experimental systems developed at various places. These used concepts such as *semantic network*, *script* and *unit* as a framework for knowledge representation.

This section concentrates not on the research history, but on methods of expressing structured knowledge, and methods of using such representation. Initially a method of representing fragmentary knowledge is described, and then its structuring and various methods of using it are discussed.

7.4.1 Unit expressions

Now we assume that the following is known:

(1) John's birthday party is being held on August 3rd.

(2) Mary is invited to the party.

(3) The party is being held at John's house.

(4) There will be a cake at the party.

This can be represented in predicate logic as follows:

(1) BIRTHDAY-PARTY(JOHN'S-B-P)
DATE(JOHN'S-B-P, AUGUST-3)
(2) GUEST(JOHN'S-B-P, MARY)
(3) PLACE(JOHN'S-B-P, JOHN'S-HOME)
(4) FOOD(JOHN'S-B-P, CAKE)

There are many advantages in collecting together these facts relating to John's birthday party, and in KRL (for Knowledge Representation Language), these collections of related facts are called units. The unit expression for John's birthday party is shown as follows:

JOHN'S-B-P
self: (ELEMENT-OF BIRTHDAY-PARTY)
date: AUGUST-3
guest: MARY
place: JOHN'S-HOME
food: CAKE

The term *slot* is applied to self (referring to JOHN'S-B-P), date, guest etc., and the descriptions written to the right of these are termed slot values.

Unit expressions can also be constructed for sets. For example, the statement:

(5) The canary is a bird

can be expressed as follows:

CANARY
self: (SUBSET-OF BIRD)

That is to say, the term canary represents the set of all canaries, which is a subset of the set of birds. Facts relating to all canaries can be expressed. For example:

(6) The colour of a canary is yellow.
(7) A canary eats grain.

Using the notation $x|$CANARY to represent $\forall x$ CANARY (x), this is shown as:

$x|$CANARY
colour: YELLOW
food: GRAIN

Unit expressions can also include variables. For example:

(8) All canaries have a human owner.

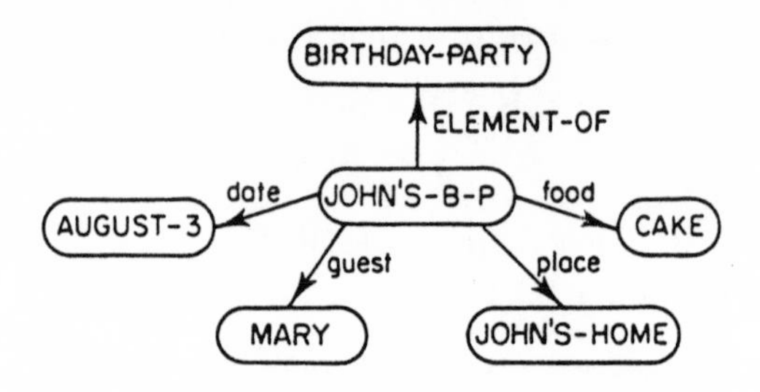

Fig. 74. Semantic network for John's birthday party

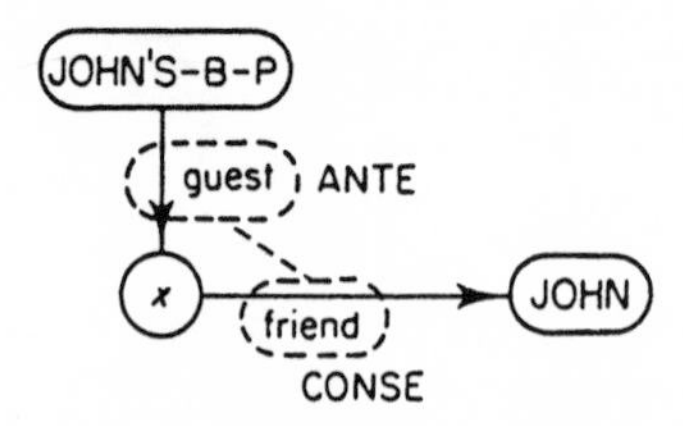

Fig. 75. Illustrative network

can be expressed as follows:

$x|$CANARY
  colour: YELLOW
  food: GRAIN
  owner: $(f(x)|$HUMAN$)$

where $f(x)$ is a Skolem function (see Section 5.3.2) and $(f(x)|$HUMAN$)$ means $\exists y$ HUMAN(y).

7.4.2 Expression by semantic networks

Unit expressions can be expressed by semantic networks; the unit for John's birthday party described above gives the semantic network shown in Fig. 74

Consider now a slightly more complicated proposition:

(9) The people invited to the birthday party are John's friends.

This proposition can be shown as follows:

$$\forall x \{\text{GUEST}(x, \text{JOHN'S-B-P}) \Rightarrow \text{FRIEND}(x, \text{JOHN})\}$$

Writing the antecedent and consequent of this proposition as respectively ANTE and CONSE, the semantic network is shown in Fig. 75

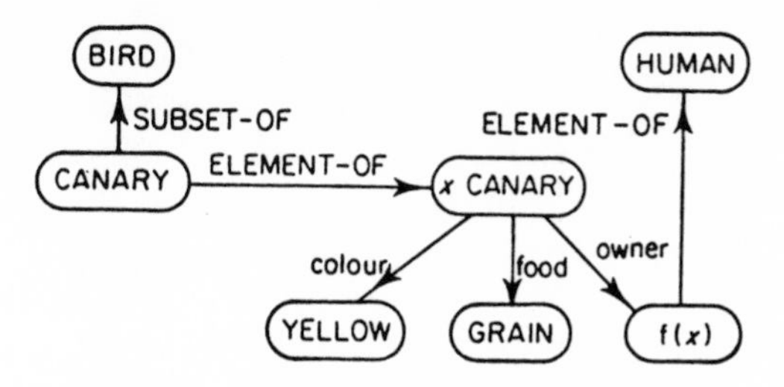

Fig. 76. Semantic network for canaries

The unit concerning canaries discussed above includes $x|$CANARY, but this is expressed as the element CANARY in a semantic network, as shown in Fig. 76.Note also that the Skolem function is shown as a node. Thus networks are useful for expressing a wide range of relationships.

7.4.3 Extraction of information from a unit expression

Suppose that the factual unit for John's birthday party shown in Fig. 74 is present, and suppose that the following question is asked: 'Is Mary invited to John's birthday party on August 3rd?' This question is represented by the following target unit:

JOHN'S-B-P
  self: (ELEMENT-OF BIRTHDAY-PARTY)
  date: AUGUST-3
  guest: MARY

This goal unit does not contradict the factual units of Fig. 74. The latter contains more information, but as long as there is no conflict, the goal unit agrees with the factual unit. If the question were 'Is Mary invited to John's birthday party which his mother is putting on?', then the target unit would include information not in the factual unit. Therefore, since this goal unit would not agree with the factual unit, the answer would be unclear. Of course, to 'Is William invited on John's birthday, August 3rd?' a negative reply would be given.

Logical deductions can also be carried out using unit expressions. The rules used in logic are expressed in a unit like that of Fig. 75. Suppose now that the question 'Who is John's friend?' is given. The target unit is:

x
  friend-of: JOHN

This agrees with the consequent in Fig. 75. Therefore the antecedent in the same diagram may be adopted as a new goal. This agrees with Fig. 74, and x is seen to be Mary.

7.4.4 Property inheritance

We have already described how the relationships between sets and members, and between sets and subsets, may be shown by unit expressions. If these relationships are used, then instead of properties of low-level units being enumerated, they can be drawn from higher-level units. This is termed *property inheritance*. Now, in addition to the canary unit, suppose that the following facts are also available:

(10) Birds can fly.
(11) Polly is a canary.
(12) Polly eats bread.

If these are added to Fig. 76, the hierarchical semantic network of Fig. 77is obtained.

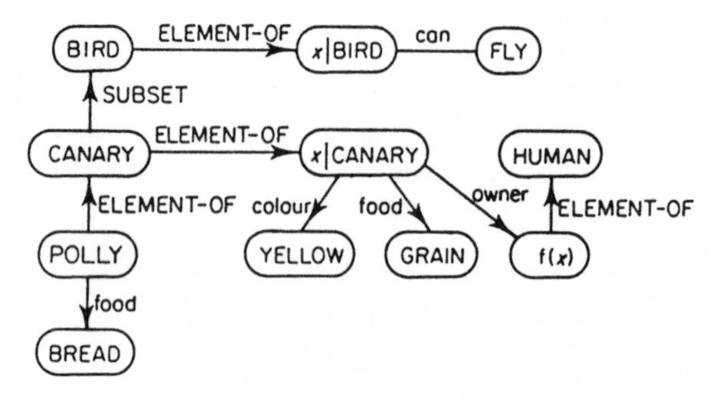

Fig. 77. Hierarchical network

If we now want to find out what food Polly eats, we inspect the food slot for Polly and find BREAD. To find Polly's colour, we first look to see if Polly has a colour slot, and if none is present we look above Polly and inspect the properties of the higher CANARY unit. In this case YELLOW is obtained. Finally, if we want to know whether Polly can fly, since this cannot be known at the level of CANARY, an answer is obtained by moving to a higher level and inspecting the properties of BIRD. By writing properties in as high-level a unit as possible, the need to repeat them for many low-level units is avoided. Provided only that special case information, such as that Polly eats bread, is written in the lower unit, correct information can always be extracted. Naturally, if there were no food slot for Polly, then the canary characteristics would be inherited.

7.4.5 Procedural attachment

Procedural knowledge can be attached to a unit. In an amended form of the unit for John's birthday party, the date slot will be written:

date: DAY 125

where DAY 125 is the name of another unit, expressed as:

```
DAY 125
  self: (ELEMENT-OF DAY)
  year: 1981
  month: 8
  day: 3
  day of week: MONDAY
```

In general, the slot value for the DAY unit may not be given in full. Also, when filling the slot with information extracted from a sentence, it is necessary to prevent incorrect values from being inserted. The following DAY unit is provided for this purpose:

```
x|DAY
  year: (y|INTEGER)
  month: (when-filled (check MONTH))
  day: (when-filled (check-day))
  day-of-week: (to-fill(get-day-of-week))
```

The year slot requires an integer. The month slot, if a value is specified, requires a check that the value is really a number (from 1 to 12) which represents a month. In the day slot, once the year and month are known, there is a call to a procedure to check that the specified date is not a contradiction, such as 29 February 1981. In the final slot there is a call to a procedure get-day-of-week, so that if the value is required it is calculated from the date. Of course, if a slot value is supplied, then that value can be used.

Thus a procedure can be attached to a unit, and another unit or program referencing this unit need not know whether or not there is a procedure in it. Attaching a procedure in this way is known as *procedural attachment*, and this technique allows the method of expression to concentrate on the overall structure of the knowledge, without making the distinction between procedural and declarative knowledge.

7.4.6 Frame description language

In the same way that KRL, described above, has been developed for units, for frames there is a language called FRL (for Frame Representation Language). This is simply an extension to LISP, providing a function to handle *frames*, which are collections of the same sort of knowledge as units. By looking here at an outline of FRL, we will understand what kind of manipulations are involved in unit-type data structures.

The frame we handle here consists of a frame name and slots. For example, the frame for the canary Polly above is represented:

```
(POLLY
  (AKO (VALUE (CANARY)))
  (FOOD (VALUE (BREAD))))
```

Here AKO stands for 'a kind of', and has the same meaning as ELEMENT-OF above.

FASSERT (names of functions manipulating frames begin with F) is used to declare a frame:

```
(FASSERT POLLY
  (AKO (VALUE (CANARY)))
  (FOOD (VALUE (BREAD))))
```

In order to add new information to this frame, the function FPUT is used; suppose that the following commands are executed:

```
(FPUT 'POLLY 'FOOD 'VALUE 'ORANGE)
(FPUT 'POLLY 'OWNER 'VALUE 'JOHN)
```

As a result the following frame will be constructed:

```
(POLLY
  (AKO (VALUE (CANARY)))
  (FOOD (VALUE (BREAD) (ORANGE)))
  (OWNER (VALUE (JOHN))))
```

To extract values from a frame, FGET is used:

```
(FGET 'POLLY 'FOOD)
*(BREAD ORANGE)
```

Similarly, FREMOVE is used to delete a slot or slot value from the frame.

In addition to assigning a value to a slot, it is also possible to assign a procedure. For example, the day-of-week slot in the DAY unit above can be assigned as follows:

```
(FPUT 'DAY 'DAY-OF-WEEK 'IF-NEEDED
'GET-DAY-OF-WEEK)
```

The function GET-DAY-OF-WEEK is then separately defined. If now

```
(FGET 'DAY 'DAY-OF-WEEK)
```

is input, first the DAY-OF-WEEK slot is tested to see if it has a value under VALUE, and if not, the function written under IF-NEEDED is evaluated.

A slot can also be given a default value. For example, the following indicates that if the year is not specified, it will be taken as 1981:

```
(FPUT 'DAY 'YEAR 'DEFAULT '1981)
```

Since, moreover, a frame can be constructed hierarchically using AKO, inheritance of higher-level characteristics is automatically achieved:

(FGET 'POLLY 'COLOUR)
*YELLOW

In summary, there are four ways to acquire a slot value:

(1) explicitly written under VALUE;
(2) evaluation of a function written under IF-NEEDED;
(3) taking a value written under DEFAULT;
(4) taking a slot value from a higher-level frame written under AKO

In practice, in FRL, to obtain information from a frame, sometimes (1) only is used, and at other times (1), (2), (4) and (3) are tried in that order.

7.4.7 Using frames

Up to this point we have discussed methods of representing collections of knowledge in terms of units and frames, and have presented an overview of what these allow us to do. The objective of Minsky's frame theory was to explain human thought patterns, and the scope for application of frames is more far-reaching than the limited description above. Without going into details of the method of implementation, we shall now discuss some of the possibilities for frames.

Imagine now a meal in a typical American restaurant. We know some things about restaurants: for example, 'The atmosphere is quiet and relaxed', and 'It takes time to have a meal', 'There is a wide choice of food', and 'It costs money'. In addition, some knowledge about 'What happens when we go into a restaurant?' is required. We might, for example, expect the following chain of events:

{ A restaurant in America}
(1) We wait for a waiter at the entrance
(2) The waiter shows us to our table.
(3) The waiter brings the menu.
(4) We choose drinks and food, and order.
(5) The food we ordered is brought.
(6) We eat.
(7) We leave a tip on the table.
(8) We pay the bill.
(9) We leave the restaurant.

Suppose the above sequence is also written in the restaurant frame. In an ordinary story it is assumed that the reader has this kind of common-sense knowledge, so that in describing a visit to a restaurant the above sequence will never be listed in full. Suppose now we are given this story:

> In San Francisco, John takes Mary to a restaurant.
> Waiter: 'Table for two, Sir?'
> 'Yes. With a good view if possible, please.'
> They sit at a table with a view of the sea.
> John says 'Choose anything you like.'
> Mary orders a steak.
> John says 'They don't keep you waiting long in this restaurant.'
> They have a pleasant meal looking out on the sea.
> As usual they tip on the generous side.
> As they leave the restaurant, John realizes that he has only 10 dollars left in his wallet.

With the first line the American restaurant frame is called. Then, expecting (1) we read the next sentence, and since there is a conversation with the waiter we understand that this is a conversation at the entrance. Next we realize that the mention of a good view is a reference to (2), and when we read 'Choose anything...' we presume (3) has ended and (4) has been reached. Continuing this process, we realize that (5) and (6) have been completed when the tip is left, and we can also understand why there are only 10 dollars left in John's wallet.

Thus, to express a collection of knowledge, frames are more efficient than the demons discussed in Section 7.2 above. Frames, what is more, allow us to change to a different frame if an expected event does not materialize.

Suppose now we enter what we believe to be a formal restaurant. If there is no-one at the entrance, and we find our own seats, it may in fact be a rather simpler establishment. If, however, we enter and find a queue of people, then we assume it may be a cafeteria. A frame may contain information on what to do if our expectations are not fulfilled; or alternatively there may be links between similar frames, so that by tracing those links the correct frame may be selected. In either case there are still many problems to be solved before this type of system can be implemented: when an expectation is not met, what cases should be considered; how should similarities and differences be defined; and can the desired frame be selected by tracing suitable links?

Chapter 8

Towards Human Intelligence

With advances in artificial intelligence research, problems of a certain level can be solved. These range, as described earlier, from simple problems in which a robot manipulates building blocks, to problems such as medical diagnosis in which expert knowledge is required. They do not, however, go so far as to replace multi-faceted human intelligence, it must be capable of solving many different problems, some of which have been discussed in previous chapters. This chapter considers some of these problems together, and while looking at a part of the phenomenon of artificial intelligence, will be reference material for readers beginning the study of artificial intelligence.

8.1 THE FRAME PROBLEM

Up to now, when a robot solves a given problem by acting on the outside world, the robot has been treated as able to carry out a number of operators, and each operator has been defined in terms of preconditions, a delete list, and an add list. Real problems cannot necessarily, however, be treated with such simple operators.

For example, consider PUTON(A, B) which puts block A on top of block B. Here we assume that, if there is space on top of a block, an unlimited number of blocks can be placed on it. If we write SPACE(y, x) to indicate that there is space on top of block y to place block x, then the preconditions for PUTON(x, y) are:

HOLD(x)
SPACE(y, x)

The delete list is not, however, obvious. Whether SPACE(y, x) is deleted or remains is determined by the upper surface area of y and whether anything is present on top of y. The add list we would expect to be ON(x, y); and certainly, in Fig. 78(a), ON(A, B) is achieved by PUTON(A, B). In Fig. 78(b), on the other hand, at the instant ON(A, B) is achieved the block structure becomes

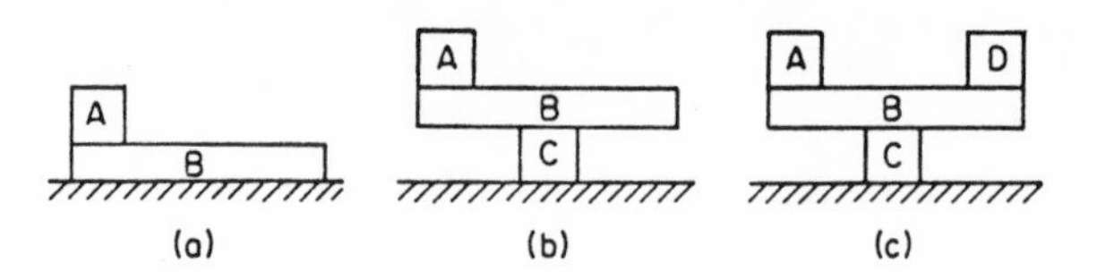

Fig. 78. When block stability becomes a problem

unstable and collapses. This does not happen if A is put in the centre of B, but it is sometimes desired to put two blocks spaced apart on top of B. In Fig. 78(c), furthermore, if block B is on the table this objective can be achieved. In the case shown in Fig. 78(c) the completed block structure is stable, but the structure shown in Fig. 78(b) occurs as an intermediate stage. Similarly, if TAKEOFF(A, B) is applied to Fig. 78(c), the remaining blocks are unstable and collapse.

The problem of what changes when an operator is applied is called the *frame problem*. This 'frame' is different from that described previously; it corresponds to the problem in an animation film of what changes from one frame to the next. In some cases only the object under consideration will move, while in others even the background will move.

If the frame problem is treated in the same way as before, an extremely large number of cases must be considered, and the operators must be defined so that no contradiction is produced. In order to achieve this, in the block problem the physical laws governing the stability of block structures would also need to be taken into account. There is a danger that if excessively detailed definitions are used the effort required to solve the problem will be too large.

One method of doing this might be the hierarchical plan described in Chapter 5; after establishing a large plan, the technique is to investigate its stability. If as a result an instability is found, the method should report back to a higher level the reason for the instability, ways of overcoming the instability, and so forth.

Then there are cases where the object world changes other than as a result of application of an operator. When filling a bath, the only operators are turning the taps on and off; but even when no operator is applied the amount of water may increase or the temperature rise. In other cases there may be several robots, or the robot may be co-existing with humans. A general method for handling these cases will not be discussed here.

8.2 *REASONING USING COMMON-SENSE*

In Chapter 5 the fact that nothing was present on top of block A was shown as:

CLEAR(A)

Let us express the present problem as:

ON(A, B)
ON(B, C)

From these expressions we would assume that block A is on top of block B, and block B on top of C, in the manner of a tower. If there were only these three blocks on the table, we would deduce CLEAR(A), although CLEAR(A) itself does not appear in the expressions.

Now, using PLANNER, described in Chapter 6, we define the following procedure:

```
(THCONSE(X Y)(CLEAR ?X)
  (THNOT (THGOAL (ON ?Y ?X))))
  ;THNOT indicates negation.
```

In other words, if there is no Y which satisfies ON(Y, X), then CLEAR(X) is established. For example, if:

(THGOAL (CLEAR B))

is input, then since (ON A B) is present in the database, it is clear that the response will be NIL. (In PLANNER, ON(A,B) is represented as (ON A B).) However, in response to

(THGOAL (CLEAR A))

the response cannot be obtained immediately. To overcome this problem in PLANNER, in response to THNOT(P) an attempt is made to prove P, and if this fails, THNOT(P) is assigned the value 'true'.

This is a method of *default reasoning*, in which elementary propositions not defined as true in the state description are assumed to be false. In ordinary logic, by the inference rules of first-order predicate calculus, if a set of propositions C can be derived from a set of non-contradictory propositions S, then even if a set S_1 of propositions which does not conflict with S is added, C can still be derived. It may, of course, be possible to derive a set C_1 of propositions additional to the set C when S_1 is added. We indicate the set of propositions derivable from S by $T(S)$ (where $T(S) \supseteq S$). Then:

$$S \subseteq S' \Rightarrow T(S) \subseteq T(S')$$

holds. In other words, T is monotonic.

In constrast, in the example above, although from the initial state description S (CLEAR A) can be derived, if ON(D, A) is added to S then (CLEAR A) can no longer be derived. Therefore, deduction using THNOT is no longer monotonic, and this type of reasoning is called *non-monotonic logic*.

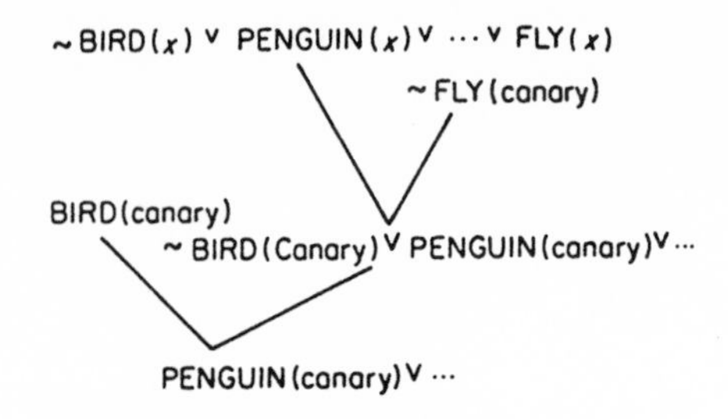

Fig. 79. Stages in the proof of FLY-(canary)

In reasoning using common-sense, humans sometimes make implicit inferences, and research has been carried out into non-monotonic logic for formal manipulation of this type of inference.

To give one more example, in common sense 'Birds can fly' is true. There are, however, exceptions such as a penguin or an ostrich. If we attempt to express this in first-order predicate calculus, we get:

$$\forall x \, \{\mathrm{BIRD}(x) \wedge {\sim}\mathrm{PENGUIN}(x) \wedge {\sim}\mathrm{OSTRICH}(x) \wedge \ldots \Rightarrow \mathrm{FLY}(x)\}$$

Now given BIRD(canary), to attempt to prove FLY(canary) using the resolution principle, a derivation such as shown in Fig. 79 is made, but then no further progress can be made. Intuitively ~PENGUIN(canary) and ~OSTRICH(canary) hold; but since they are not given directly, the empty clause cannot be derived. In this case again, if since the elementary proposition PENGUIN(canary) is not given we assume that its inverse holds, then FLY(canary) is proved.

It is inconvenient to write the above long expression in order to express 'Normally birds can fly', and so we introduce the symbol M to play a role similar to that of THNOT. If M is appended to FLY(x), then to say M FLY(x) is true is to indicate that ~FLY(x) cannot be derived. The meaning of M FLY(x) ⇒ FLY(x) is that, if ~FLY(x) is not proved, then FLY(x). The above form will here also be called a *proposition*. Introducing this symbolism, the above relationship relating to birds will be written:

$$\mathrm{BIRD}(x) \wedge \mathrm{M\ FLY}(x) \Rightarrow \mathrm{FLY}(x)$$

We deduce that FLY(canary) is true from BIRD(canary) and the absence of ~FLY(canary).

Consider, however, a set S containing in addition to propositions of first-order predicate calculus the following:

$$\mathrm{M}\ P \Rightarrow {\sim}Q$$
$$\mathrm{M}\ Q \Rightarrow {\sim}P$$

Further suppose that S does not contain any of P, $\sim P$,Q, or $\sim Q$. If we first apply the first of these expressions, then $\sim Q$ holds, and then of course $\sim P$ does not hold. If the second expression is applied first then $\sim P$ holds and $\sim Q$ does not hold. There are therefore two types of sets of propositions derivable from S, one of which includes $\sim Q$ and excludes $\sim P$. It is sometimes not possible to determine a single set of propositions derived in this way. If we suppose that S is a set including:

$$\mathrm{M}\ P \Rightarrow \sim P$$

then if $\sim P$ holds, from the expression, $\sim P$ no longer holds. But if P holds, then again $\sim P$ holds. This example contains an intuitively unnatural proposition, and such problematical propositions must be eliminated from sets of propositions. Generally, given a large number of propositions it is not simple to find the propositon giving the problem.

In order to express the fact that birds can fly, in the units and frames described in Chapter 7, we use a hierarchical structure. In other words, in the bird unit we write that birds can fly, and in the penguin unit we write that penguins cannot fly. Then to find the value of FLY(x) we first look in the x unit. If we find 'can fly' or 'cannot fly' there, then we need look no further; but if it is not given we refer to a higher unit in the hierarchy. For example, if nothing is written in the CANARY unit then we refer to the bird unit, and find that it can fly.

This method is effective for things which have a well-behaved hierarchy. There may, however, be several different hierarchies for the same objects, depending on the viewpoint, and the problem becomes one of knowing which hierarchical structure to investigate. There again, for a natural language statement such as 'If a person gets in a boat, he can move it. (There are no holes in the boat, there are oars, and the person has the strength to row.)' the problem cannot be solved simply by constructing a hierarchy.

When, after reasoning using common-sense, new facts are discovered, the results of deductions made up to that point must be amended. This is called *truth maintenance* and is the subject of present research. Implicit reasoning and truth maintenance on a human level will be very important in constructing question and answer systems.

8.3 KNOWLEDGE ACQUISITION

To solve complicated problems a large amount of knowledge is necessary, and this knowledge must be expressed in a suitable format, as already discussed. The problem is, however, in what way a large amount of knowledge should be presented to the system. To input the state and rules of the universe in question as a computer program requires a considerable effort, and knowledge acquisition is thus an important topic in artificial intelligence.

The approaches currently being considered are systems to make the input of knowledge simpler (metasystems), and knowledge acquisition by learning.

In the field of knowledge engineering, the development is the advancement of metasystems in which the knowledge necessary for an expert system is input directly by a specialist in the relevant field. Conventionally, computer specialists would listen to the explanation of a specialist in the field, and translate that into a suitable form for inputting to the computer. Generally, in the systems being developed, the method of knowledge representation and the inference mechanism are fixed and individual items of knowledge are obtained from an expert using a restricted natural language. EMYCIN, for example, constructs a system having functions similar to MYCIN, and preserves the following characteristics of MYCIN:

(1) backwards deduction based on production rules;
(2) control of searching and evaluation by certainty factors;
(3) interrogation in English;
(4) functions to explain the system's stages of reasoning.

If specialist knowledge corresponding to that of MYCIN is input to EMYCIN, then the desired system is completed. The input of rules in MYCIN can be done in natural language; but in EMYCIN, in place of long natural language sentences an ALGOL-like language can be used. EMYCIN converts these into an internal LISP type of expression. These internal representations can be converted to English to confirm the rules input. A simple example is shown in Table 12. EMYCIN also has the ability to respond to English questions based on the internal expressions. It can also, given a number of examples, evaluate how effectively a given set of rules can be applied to those examples.

In addition to EMYCIN, other metasystems have been constructed, such as Unit Package which expresses knowledge using units, and AGE, which includes the hierarchy of the production rules, and in which the order of rule application can be controlled.

Expert systems must store knowledge which has not been formalized, and in a large amount of knowledge there may therefore be contradictions. Furthermore, a rule which holds for one case may possibly produce a wrong decision for a different case which was not considered when the rules were input. As the system gets larger it becomes increasingly difficult to investigate contradictions *a priori* and to eliminate obvious errors, so that once the system is initially set up many improvements are necessary. These problems must also wait for future research.

Investigation have also been carried out into direct knowledge acquisition by learning since the early days of artificial intelligence; initially most methods simply made parameter adjustments. The well-known draughts-playing program by A. L. Samuel used as its evaluation function for determining the

TABLE 12
Rule input using EMYCIN

(a) Rule input
If composition = (LISTOF METALS) and
Nd-stress >.5 and
Cycles >1000
Then Ss-stress = fatigue

(b) Internal representation
PREMISE: ($AND
(SAME CNTXT COMPOSITION (LISTOF METALS))
(GREATERP* (VAL1 CNTXT ND-STRESS) .5)
(GREATERP* (VAL1 CNTXT CYCLES) 10000))
ACTION: (CONCLUDE CNTXT SS-STRESS FATIGUE TALLY 1.0)

(c) English expression
If: (1) the material composing the substructure is one of the metals,
(2) the non-dimensional stress of the substructure is greater than 0.5, and
(3) the number of cycles the loading is to be applied is greater than 10 000
Then: it is definite (1.0) that the fatigue is one of the stress behaviour phenomena in the substructure.

next move a weighted sum of functions reflecting various strategies, and adjusted the weightings according to whether the program won or lost. Pattern classification systems using learning attach a weight to each picture element of a binary-valued pattern, and classify a pattern according to those values. If a number of known patterns are supplied, and the system is shown to what category each belongs, it can change the weightings suitably according as each picture element is 0 or 1. After this sort of learning stage, both known patterns and similar unknown patterns can be classified largely correctly.

A system for learning about building block structures was constructed by P. H. Winston at MIT in 1970. If shown examples and counter-examples of a structure by a human, this system produces correct descriptions. Fig. 80 shows examples and ('near miss') counter-examples used to explain the meaning of 'arch'. The system first constructs from Fig. 80(a) the description shown in Fig. 81(a), which has nodes and labelled arrows called pointers joining them. The figure indicates that the arch shown in constructed from elements A, B, and C (ONE-PART-IS pointers), that each is a cuboid (BRICK: A-KIND-OF pointer), A is supported by B and C (SUPPORTED-BY pointers), B is left of C (LEFT-OF), C is right of B (RIGHT-OF), and that they do not touch (DO-NOT-TOUCH). This is thus a symbolic representation of Fig. 80(a). Next, presented with the near miss of Fig. 80(b), the system builds the description shown in Fig. 81(b). Comparing this with the first arch, the correspondences between the elements are determined, and then the pointers are compared. In this case, it is found that SUPPORTED-BY has changed to

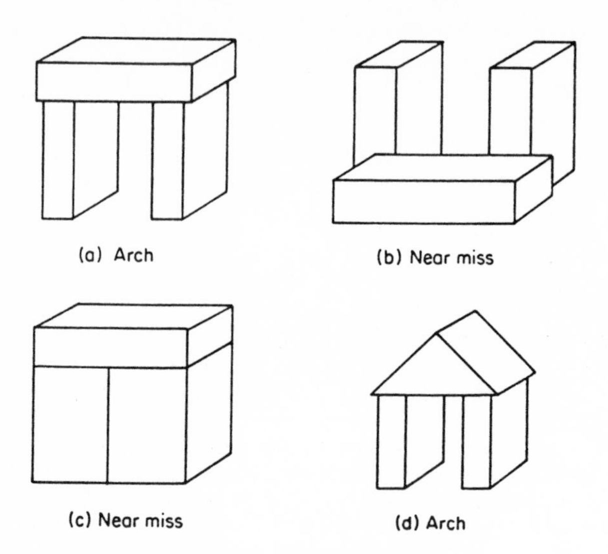

Fig. 80. Scenes for learning 'arch': (a) arch, (b) near miss, (c) near miss, and (d) arch

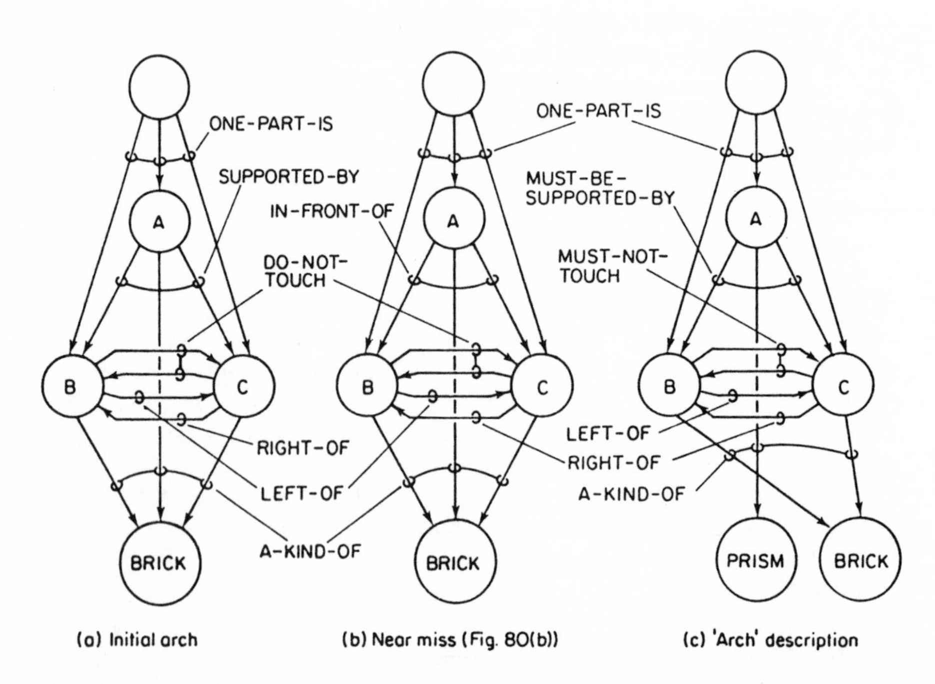

Fig. 81. Stages in learning 'arch': (a) initial arch, (b) near miss, and (c) 'arch' description

IN-FRONT-OF. Thus it is seen that SUPPORTED-BY is a necessary condition for an arch, and the SUPPORTED-BY pointer in the first arch description is modified to MUST-BE-SUPPORTED-BY. Presented next with Fig. 80(c), the system determines, in a similar way, that DO-NOT-TOUCH is a necessary condition, and this pointer is amended to MUST-NOT-TOUCH. Finally the correct arch of Fig. 80(d) is presented. When the arch description produced so far is compared with the new description, it is realized that the A-KIND-OF pointer for A no longer points to a BRICK, but to a WEDGE, and it can be understood that A can be either a BRICK or a WEDGE. The modification here can be made in several ways; for A any of 'either BRICK or WEDGE', 'right prism', 'prism', 'polyhedron' or 'any block' leads to no contradiction. Fig. 81(c) shows the modified result when PRISM is used.

In this learning process pointers and the destination of pointers are amended. Changing pointers is a process similar to learning by changing weightings, but a pointer is different in being one of a finite set of symbols, rather than a simple number like a weighting. The latter process is one of generalizing the object which is the target of a pointer by moving up the hierarchical tree classification of the objects.

Some research has also been done into the generalization problem: given a number of descriptions, produce a description which fits them all. The following are some simple rules for generalization:

(1) Eliminate part of the descriptions. For example, a generalization of RED(x) v BIG(x) and RED(x) v SMALL(x) is RED(x).
(2) Change a constant to a variable. For example, obtain ON(x, block1) from ON(pyramid1, block1) and ON(block2, block1).
(3) Set a range. For example, for HEIGHT(block1) $= a$ and HEIGHT(block2) $= b$, derive HEIGHT(x) in between a and b.
(4) Move up the classification tree (the method used in the arch example).

There have been some experiments with expert systems which learn production rules. It is then not necessary for the rules to hold for all cases. For example, a system Meta-DENDRAL, which learns the rules for DENDRAL (described in Chapter 7), generalizes rules to hold for most cases, even if there are small exceptions. Specifically, mass spectra of materials of known structure are input as data, and for each material the ways in which it can be decomposed are enumerated. Then a large number of rules are produced showing where the decomposition occurs. Finally, eliminating redundant rules (those which are a special case of another rule), rules with exceptions are specialized, and those which can be are generalized. Even rules which humans have not noticed have been discovered this way.

Apart from Meta-DENDRAL there have so far been few examples of rules found which humans had been unable to find, and it cannot be said that the learning method is entirely satisfactory. For example, in systems learning with

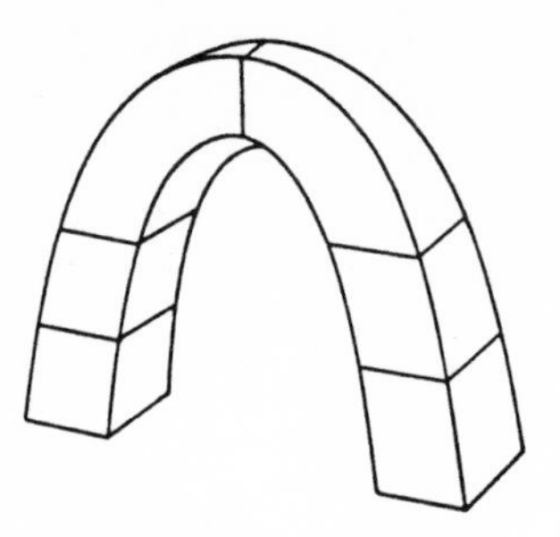

Fig. 82. Another arch

building blocks, the system may well have problems if Fig. 82 is shown as an arch. In order to regard this also as an arch, it is necessary to describe an arch as a structure like a short tunnel, and for this the system must first construct this description when presented initially with Fig. 80(a). It is difficult to enlarge the description from that constructed so far when presented with an example at an intermediate stage.

Whatever learning technique is adopted, the construction of a long description is the key to successful learning. The system is not told beforehand what is required to describe the object of learning, and many problems remain for the creation of a system which can, like a human, learn from experience without having a detailed specification of how to produce descriptions.

8.4 KNOWLEDGE OF THE EXTERNAL WORLD

As described under knowledge acquisition, for a system to obtain knowledge, a human inputs information in a form suited to the system. At the present level of artificial intelligence, even when learning by being shown examples, it will be difficult for the system to provide a suitable description when shown a complicated object.

Character recognition techniques do not allow an ordinary book to be read as it is, and reading sentences in ordinary handwriting is a problem for the future.

Devices are available which can recognize a limited vocabulary of speech, but these have not reached the stage of capturing a radio news bulletin or primary school lesson, for example.

It might be said that so-called pattern information processing is lagging when compared with symbol manipulation or logical inference. If it once reaches a certain level, however, the necessary conditions for learning in the same way as a human will be fulfilled. There will no longer be a need to design special interfaces between human and computer, and knowledge acquisition may be expected to become extremely easy. Carrying out this kind of pattern information processing will require not only the development of various methods of signal processing, but also an effective method of using knowledge of the world. It seems that there is a large body of common knowledge required to recognize written sentences, the spoken word, drawings, man-made objects, or a natural scene. If this common knowledge can be converted into a database in a suitable form, this will no doubt aid the development of systems for such tasks as natural language processing, speech comprehension, picture recognition and image analysis.

Bibliography

CHAPTERS 1–4: ARTIFICIAL INTELLIGENCE IN GENERAL

1. Feigenbaum, E. and Feldman, J. (eds.), *Computers and Thought*, McGraw-Hill, 1963.

   A collection of papers from early artificial intelligence research. Deals principally with automatic theorem-proving, a learning draughts-playing system, and GPS.

2. Minsky, M. (ed.), *Semantic Information Processing*, MIT Press, 1968.

   A collection of MIT artificial intelligence research papers. Detailed discussion of well-known research on semantic networks, natural language processing, recognition of geometrical similarity, etc.

3. Nilsson, N. J., *Problem Solving Methods in Artificial Intelligence*, McGraw-Hill, 1971.

   A clear exposition focusing on problem-solving by means of searching and predicate calculus.

4. Winston, P. H. (ed.), *The Psychology of Computer Vision*, McGraw-Hill, 1975.

   A collection of research papers from MIT relating to vision. Articles on line-drawing interpretation, image processing, learning and frame theory.

5. Winston, P. H., *Artificial Intelligence*, Addison-Wesley, 1977.

   An introductory text aimed at early undergraduate years. Early part of book discusses main fields of artificial intelligence, centring on research at MIT. Later part deals with LISP and its application to artificial intelligence research.

6. *Information Processing*, **10**, (10), 1978: special issue on artificial intelligence and software techniques.

   Twelve topics, none more than ten pages, on the prospects for artificial intelligence and a wide range of related subjects (in Japanese).

CHAPTER 5

7. Ernst, G. W. and Newell, W., *GPS: A Case Study in Generality and Problem Solving*, Academic Press, 1969.

   A detailed exposition of GPS, with many examples of application to cannibal and missionary and integration problems.

8. Nagao, S. and Fuchi, K., *Logic and Meaning*, Iwanami Shoten.

A more detailed treatment than in this book of predicate calculus and the resolution principle (in Japanese).

9. Fikes, R. E., Hart, P. E. and Nilsson, N. J., 'Learning and executing generalized robot plans', *Artificial Intelligence*, **3**, 1972, 251–288.

Proposes a method for robot planning, using triangle tables.

10. Sacerdoti, E. D., 'Planning in a hierarchy of abstraction spaces', *Artificial Intelligence*, **5**, 1974, 115–135.

Proposes hierarchical planning, with examples.

CHAPTER 6

11. McCarthy, J., Abrahams, P. W., Edwards, D. J., Hart, T. P. and Levin, M. I., *LISP 1.5 Programmers' Manual*, MIT Press, 1962.

LISP introduction by its creators. All LISP systems are based on this book. Rather academic in parts.

12. Weissman, C., *LISP 1.5 Primer*, Dickenson, 1967.

A good introduction.

13. Nakanishi, M., *Introduction to LISP*, Kindai Kagakusha, 1977.

Reference for creating a LISP system (in Japanese).

14. Winston, P. H. and Horn, B. K. P., *LISP*, Addison-Wesley, 1981. A detailed reference work with many examples and exercises. Readable discussion from an introduction to LISP to problem-solving natural-language processing, etc. Also deals with creating a LISP interpreter or compiler.

15. Hewitt, C., *Description and Theoretical Analysis of PLANNER,* AI TR-258, MIT AI Lab., 1972.

Proposal for PLANNER, though not the implemented version.

16. Sussman, C. J. and Winograd, T., *MICRO-PLANNER Reference Manual*, AI Memo, No. 203, MIT AI Lab., 1970.

Manual for implemented version of MICRO-PLANNER. Similar discussion also in: Winograd, T., *Understanding Natural Language*, Academic Press, 1972.

17. Sussman, G. J. and McDemott, D. V., 'From PLANNER to CONNIVER', *Proc. FICC*, 1972, 1171–1180.

Describes features of CONNIVER.

18. Yokoi, T., *Information Processing*, **17**, 1976, 577–586; *ibid.*, **17**, 1976, 760–768.

Discussions of current artificial intelligence languages concentrating on GPS, PLANNER, and CONNIVER (in Japanese).

19. Pereira, L. M., Pereira, F. C. N. and Warren, D. H. D., *User's Guide to DEC system 10 PROLOG,* Dept. AI Research, Edinburgh Univ., 1978.

Manual for PROLOG, developed at Edinburgh, including simple examples.

20. Fuchi, K., 'Predicate logic type languages', *Information Processing*, **22**, 1981, 588–591.

An easy introduction to PROLOG and its features (in Japanese).

CHAPTER 7

21. Charniak, E., *Toward a Model of Children's Story Comprehension*, Ph.D. Thesis, MIT AI Lab., 1972.

Proposes using demons to understand a story.

22. Newell, A. and Simon, H. A., *Human Problem Solving*, Prentice-Hall, 1972.

A theoretical treatise from the proponents of production systems.

23. Feigenbaum, E. A., Buchanan, B. G. and Leiderberg, J., 'On generality and problem solving: a case study using the DENDRAL program', *Machine Intelligence*, **6**, 1971.

An overview of DENDRAL and examples of rules used.

24. Shortliffe, E., *Computer-Based Medical Consultations: MYCIN*, American Elsevier, 1976.

A detailed treatment of MYCIN.

25. Feigenbaum, E. A., 'The art of artificial intelligence: I. Themes and case studies of knowledge engineering', *Proc. 5th Int. Joint Conf. on Artificial Intelligence*, 1977, 1014–1029.

A discussion from Stanford University, where research into experts systems using production systems is the most advanced of various systems, particularly DENDRAL and MYCIN. The term 'knowledge engineering' came to be used from this time.

26. Bobrow, D. G. and Winograd, T., 'Experience with KRL-0 one cycle of a knowledge representation language', *Proc. 5th Int. Joint Conf. on Artificial Intelligence*, 1977, 213–222.

An overview of KRL, which uses units, and its pros and cons.

27. Minsky, M., 'A framework for representing knowledge', in *The Psychology of Computer Vision*, ed. P. H. Winston (see Reference 4).

Introduces frames. Discusses what structures are necessary to understand how humans see objects or listen to speech.

CHAPTER 8

28. *Artificial Intelligence*, **13**, 1980, (1): special issue on non-monotonic logic.

Six papers on the frame problem, non-monotonic logic, etc.

29. Nilsson, N. J., *Learning Machines*, McGraw-Hill, 1965.

A good explanation of typical methods of training with two-dimensional patterns to distinguish unknown patterns.

30. Winston, P. H., 'Learning structural description from examples', in *The Psychology of Computer Vision*, ed. P. H. Winston (see Reference 4).

A detailed description of a learning system using building blocks.

Index